T0139101

# Building a Dedicated GSM GPS Module Tracking System for Fleet Management

## Hardware and Software

# Building a Dedicated GSM GPS Module Tracking System for Fleet Management

## Hardware and Software

Franjieh El Khoury
Antoine Zgheib

CRC Press
Taylor & Francis Group
Boca Raton  London  New York

CRC Press is an imprint of the
Taylor & Francis Group, an **informa** business

CRC Press
Taylor & Francis Group
6000 Broken Sound Parkway NW, Suite 300
Boca Raton, FL 33487-2742

© 2018 by Taylor & Francis Group, LLC
CRC Press is an imprint of Taylor & Francis Group, an Informa business

No claim to original U.S. Government works

Printed on acid-free paper

International Standard Book Number-13: 978-1-4987-6702-6 (Hardback)

---

**Library of Congress Cataloging-in-Publication Data**

Names: El Khoury, Franjieh, author. | Zgheib, Antoine, author.
Title: Building a dedicated GSM GPS module tracking system for fleet management : hardware and software / Franjieh El Khoury, Antoine Zgheib.
Description: Boca Raton : Taylor & Francis, 2018. | Includes bibliographical references.
Identifiers: LCCN 2017044898 | ISBN 9781498767026 (hb : alk. paper)
Subjects: LCSH: Motor vehicle fleets--Equipment and supplies. | Automatic tracking. | Global Positioning System--Industrial applications.
Classification: LCC TL165 .E423 2018 | DDC 629.2/72--dc23
LC record available at https://lccn.loc.gov/2017044898

---

**Visit the Taylor & Francis Web site at**
**http://www.taylorandfrancis.com**

**and the CRC Press Web site at**
**http://www.crcpress.com**

To the people who made significant contributions to our life,

We dedicate this work.

After the rain, nice weather always ends up coming...

# Contents

# Acknowledgments

First, we would sincerely like to thank all the members of CRC Press/ Taylor & Francis Group for their patience and support, especially the Chief Editor Mr. Rich O'Hanley.

We would like to thank the open source application OpenGTS's development team for their guidance, and to Mr. Martin Flynn, in particular.

We are especially grateful to "Libelium" and the marketing department for their efforts to provide us with the necessary documentation and full support in allowing us to use their products.

The expression of our sincere gratitude goes to the members of the ERIC Research Lab, especially Professor Marcel Egea and Professor Stephane Bonnevay, for their professional encouragement.

Thanks also to Mr. Gilles Cattan, Engineer and Former Director of R&D of SFIM industries for his advice and moral support.

We sincerely thank all of our families for all the sacrifices they made so we could achieve our goal.

# Introduction

The rapid evolution of the GPS module tracking systems for fleet management using customized commercial applications as well as open source applications (e.g., OpenGTS) have increased the network's constraints in terms of usage. In the last few years, 3G modules have proven their applicability in providing efficient information to the end user (e.g., real time geolocation). Furthermore, the integration of microcontrollers has evidently been efficient in resolving several problems in GPS devices.

There were difficulties encountered from the disconnection of a network to collect the information from a satellite, the cost of using the Internet to download the data, and the non-adaptability of the open source applications (e.g., OpenGTS) with some GPS devices to manage the data and edit the required reports. These difficulties were in regard to getting accurate information about a location, with increases in the risk of losing data and the cost of developing customized application, as well as causing problems to accessing and to collecting the required data.

In order to provide the end user with accurate GPS information and guarantee the access to collecting the required data at any moment, our study focuses on the development of the model INFelecPHY GPS Unit (IEP-GPS) tracking system for fleet management that is based on 3G and general packet radio service (GPRS) modules. This model should offer reliability since it deals with several protocols: (1) HTTP and HTTPS to navigate, download and upload in real time the information to a Web server, (2) FTTP and FTTPS to handle in a non-real time the files of the Web application, and (3) SMTP and POP3 to send and receive e-mail directly from the unit in case of any alert.

This model is similar to a mobile device, but without a screen for display. It is multifunctional, since it will be linked to a GPRS module, a camera, a speaker, headphones, a keypad, and a screen. This model will provide accuracy, since it will work when we do not have a connection to the network to store on a SD card socket the necessary tracking information and will be sent automatically once connected to the Internet during the idle time.

Our proposed model IEP-GPS is compatible with the open source application, OpenGTS, to do tracking for vehicles on various types of maps and edit various reports. Also, this model solves the problem of accessing and collecting the data from the server of the provider company, when there is no Internet connection, by integrating the feature of implementation of the application on the local server of the company.

Therefore, our work focuses on the development of dedicated 3G module tracking systems and microcontrollers in order to provide greater efficiency in the IEP-GPS model. Our approach aims to improve the existing models for the storage of the information, when there is a disconnection from the network, as well as its flexibility to gather the required information for the company without the need of the Internet with less cost and in a faster time than the existing systems.

### Presentation of the Book

This book is composed of several chapters to achieve a relevant model able to provide accurate information and guarantee access for collecting the required data at any moment.

In Chapter 1, we define the Global Positioning System (GPS) and the method to locate a position by the GPS receiver. In addition, we detail the different segments of GPS, the different types of GPS receivers, and the application fields of GPS receivers.

We develop in Chapter 2 the different electronic equipment for a GPS system by providing the classification of microcontrollers and their different types with the effectiveness of each type. As well, we detail the Future Technology Devices International (FTDI), the programmer, the bootloader, the cables and connectors, the SD card, the GPS antenna, the GPS camera, and the audio kit.

In Chapter 3, we present the various communication modules with the effectiveness of each module. We also detail the different communication protocols with the effectiveness of each protocol to exchange the information with the connected external physical peripherals and the various Internet communication protocols with the effectiveness of each protocol to transmit the required data from the GPS unit to the end user.

We describe in Chapter 4 the development system of microcontroller applications and the various programming languages of the microcontrollers with the effectiveness of each language. Also, we detail the ATtention (AT) command to control the modem function, and we present the National Marine Electronics Association (NMEA) sentence and the GPS recommended minimum data (GPRMC) sentence format to convert into them the received GPS information from the satellite.

In Chapter 5, we propose our model IEP-GPS, based on the GPRS/3G module, to transmit the information from the IEP-GPS device to the end user, and the architecture of the model with and without integration of a microcontroller. Moreover, we detail the interactions between the different actors of the model for both proposed architectures of the model. Furthermore, we describe the three modules of the composition of the model: (1) obtaining a location and posting to the Web, (2) sending SMS as an alert message, and (3) sending photos by e-mail using the SMTP protocol service.

Chapter 6 is devoted to the development of the three modules of our proposed model in C programing language, Lua scripting language, and the AT command set. In addition, we illustrate a simulation on tracking the history path of a given user and showing the details about the first location when moving out of the geofence.

Finally, we conclude with a summary of all the accomplished work and recommendations that can be realized in the near future.

# 1

# GLOBAL POSITIONING SYSTEM (GPS)

## 1.1 General Overview of Global Positioning System (GPS)

According to Michael Dunn (Dunn 2013) and Joel McNamara (McNamara 2004), Global Positioning System (GPS) is known as NAVSTAR (Navigation Satellite Timing and Ranging) GPS. It provides geolocation and time information to a GPS receiver in all weather conditions. It works independently of any telephonic or Internet reception.

GPS is defined by Joel McNamara (McNamara 2004) as a smart satellite system that can pinpoint a position anywhere on planet Earth. GPS is a radio receiver measuring the distance from a given location to satellites that orbit the Earth broadcasting radio signals. The GPS receiver needs to get the location from a minimum of three satellite signals. Moreover, it requires four satellite signals to get the position in three dimensions (i.e., latitude, longitude, and elevation). GPS is used for navigation and location positioning by the military, the government, and civilians; however, radio signals have been used for navigation since the 1920s.

According to Luc Aebi (Aebi 2007), the LORAN (Long Range Navigation) system was developed during World War II to measure the time difference between sending and receiving radio signals. The LORAN system is used for maritime navigation, and it works by obtaining the position in two dimensions (i.e., latitude and longitude).

In 1957, the Russians launched the first satellite, Sputnik, to orbit the Earth by using a radio transmitter to broadcast telemetry information (McNamara 2004).

In 1959, the first satellite in radio navigation, TRANSIT, was launched using helpful and reliable technologies for the GPS system (Aebi 2007).

By the 1960s, the U.S. Army, Navy, and Air Force were all working on independent versions of radio navigation systems (i.e., primary satellite positioning systems) that could provide accurate positioning in all weather conditions and 24/7 coverage (McNamara 2004). In 1973, the Air Force consolidated all the military satellite navigation efforts into a single program, the NAVSTAR Global Positioning System, which was launched in 1974 to test the concept.

In April 1980, the first satellite GPS was launched by the U.S. Department of Defense (DoD) to make the system functional (Aebi 2007). In 1982, the DoD reduced the GPS constellation of satellites from 24 to 18 owing to budget cuts. Then in 1983, GPS was available for civil aviation use. In 1988, the GPS constellation of satellites was increased from 18 to 21 plus three spare satellites.

In 1990, the first GPS receiver for the general public was marketed by TRIMBLE (Aebi 2007). At the same time, the DoD decided to establish a known signal degradation system, SA (selective availability). This system was activated in July 1991 after the first Gulf War, but was finally removed in May 2000.

In 1994, the planned full constellation of 24 satellites was in place and the system was completely operational (McNamara 2004), as shown in Figure 1.1.

In addition to GPS, there are other systems in use or under development that have the same functionalities as GPS, such as the Russian Global Navigation Satellite System (GLONASS), the European Union Galileo positioning system, China's BeiDou Navigation Satellite System, the Japanese Quasi-Zenith Satellite System, and India's Indian Regional Navigation Satellite System (NAVIC).

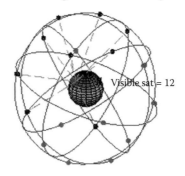

**Figure 1.1**    GPS constellation. (Source: Jan, S. 2010. GPS Segments / Components.xa.yimg.com/kq/groups/21620206/615119600/name/3.+GPS+Segments.ppt.)

## 1.2 Locating a Position with the GPS Receiver

The distance between the position of the GPS satellite and the GPS receiver is calculated by using Equation 1.1 (Corvallis Microtechnology 2000; McNamara 2004; Raju 2004; TTU 2012).

$$\text{Distance} = \text{speed} \times \text{time} \tag{1.1}$$

In other words, a GPS receiver determines the amount of time it takes the radio signal (i.e., GPS signal) to travel from the GPS satellite to the GPS receiver. The GPS signal travels at the speed of light (186 thousand miles per second). Both the GPS satellite and the GPS receiver generate an identical pseudo-random code sequence. When the GPS receiver receives this transmitted code, it determines how much the code needs to be shifted (using the Doppler-shift principle) for the two code sequences to match. Therefore, the shift is multiplied by the speed of light to determine the distance from the GPS satellite to the GPS receiver.

GPS satellites are orbiting the Earth at an altitude of 11 thousand miles (Corvallis Microtechnology 2000; Raju 2004; TTU 2012). Assuming that the GPS receiver and the satellite clocks are precisely and continually synchronized, the GPS receiver uses three satellites to triangulate a 3D position, then the GPS provides coordinates (X, Y, Z) for a calculated position. However, a GPS receiver needs four satellites to provide a 3D position, as shown in Figure 1.2. Since the GPS receiver clock is not as accurate as the atomic clocks in the satellites, then a fourth variable T for time is determined in addition to the three variables (X, Y, and Z). Moreover, the GPS signals travel from

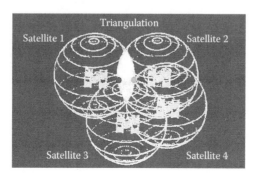

**Figure 1.2** Basic principle of positioning with GPS. (Source: TTU. 2012. Principles of GPS. In *Lectures documents*, Geospatial Center, Texas Tech University, Texas, USA.)

the GPS satellite to the GPS receiver very fast, thus, if the two clocks are off only a small fraction of time, the determined position may be inaccurate.

The DoD can predict the paths of the satellites vs. time with great accuracy. It constantly monitors the orbit of the satellites looking for deviations, known as ephemeris errors, from predicted values (Corvallis Microtechnology 2000; McNamara 2004; TTU 2012). Once these errors are detected for a given satellite, they will be sent back up to that satellite, which broadcasts them to the GPS receivers as a standard message. Nowadays, the GPS receivers store the orbit information, known as an almanac, for all the GPS satellites (Corvallis Microtechnology 2000; McNamara 2004; TTU 2012). Therefore, this information advises about the position of each satellite at a particular time. Moreover, this information in conjunction with the ephemeris error data can help to determine in a very precise way the position of a GPS satellite at a given time.

## 1.3 Segments of GPS

We distinguish three types of GPS segments (El-Rabbany 2002; Jan 2010; Neilan and Kouba 2000; Raju 2004; TTU 2012): the space segment, the control segment, and the user segment, as shown in Figure 1.3.

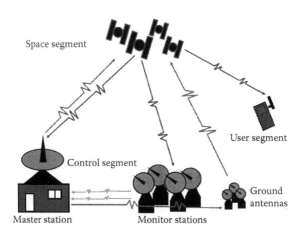

**Figure 1.3**   GPS segments. (Source: TTU. 2012. Principles of GPS. In *Lectures documents*, Geospatial Center, Texas Tech University, Texas, USA.)

### 1.3.1 Space Segment

The space segment is composed of the orbiting GPS satellites or space vehicles. Satellites orbit at an altitude of approximately 20,200 kilometers. Each space vehicle makes two complete orbits each sidereal day, repeating the same ground track. The orbits are arranged to have at least six satellites within line of sight from almost everywhere on the Earth's surface.

The space segments consist of 24 satellite constellations, eight each in three circular orbital planes. Then, the GPS satellites are equally distributed in each of the six orbit planes with four satellites each. Each GPS satellite transmits a signal. Each signal contains a number of components:

- Two sine waves known as carrier frequencies.
- Two digital codes.
- A navigation message.

The carriers and the codes are used to determine the distance from the user's receiver to the GPS satellites.

The navigation message contains the coordinates (i.e., location) of the satellite as a function of time.

The transmitted signals are controlled by highly accurate atomic clocks onboard the satellites.

### 1.3.2 Control Segment

The control segment of the GPS system is controlled by the U.S. Army. It consists of a worldwide network of tracking stations, with a Master Control Station (MCS) located in the United States.

The control segment aims to operate and monitor the GPS system. The main objective of the control segment is to track the GPS satellites in order to determine and predict satellite locations, system integrity, behavior of the satellite atomic clocks, atmospheric data, the satellite almanac, and other information. This information is uploaded into the GPS satellites through the S-band link.

The control segment is composed of:

- A Master Control Station (MCS).
- An Alternate Master Control Station.

- Four dedicated ground antennas.
- Six dedicated monitor stations.

The monitor stations are operated in the UK, Argentina, Ecuador, Bahrain, Australia, and Washington, DC. In 2005, six more monitor stations were added to the grid, which allows calculating more precise orbits and ephemeris data, as well as a better position precision for the end user.

### 1.3.3 User Segment

The user segment includes all military users (i.e., U.S. and allied military users) of the secure GPS Precise Positioning Service, all civilian users, all GPS receivers and processing software, and commercial and scientific users of the Standard Positioning Service.

With a GPS receiver connected to a GPS antenna, a user can receive the GPS signals, which can be used to determine their position anywhere in the world.

Public users apply the GPS for navigation, surveying, time and frequency transfer, and other uses.

GPS is currently available to all users worldwide at no direct charge.

### 1.4 Different Types of GPS Receivers

We distinguish five types of GPS receivers (El-Rabbany 2002; McNamara 2004; Rodgers 2007): consumer model, U.S. military/government model, mapping/resource model, survey model, and commercial transportation model.

### 1.4.1 Consumer Model

The consumer GPS receivers are easy to use and are targeted for recreational and other uses that do not require a high level of location precision. They can be found in sporting goods stores at a relatively low cost. There are three main manufacturers for the consumer GPS receivers in the market: Garmin, Lowrance, and Magellan GPS. Consumer GPS devices can be installed as additional options on handheld devices, mobile phones, smartphones, palm pockets, and laptop PCs, as well as desktop computers. Most of the converged devices include GPS phones and GPS cameras.

It is recommended to use the consumer GPS receivers with 12-channels and not 8-channels. Therefore, the consumer GPS receivers with 12-channel parallel acquire satellite signals faster and operate better under foliage and tree-canopy cover. However, the consumer GPS receivers with 8-channels acquire satellite signals slowly and hardly operate even under light tree cover. The consumer model can be between 15 to 30 meters accurate.

We distinguish two types of consumer GPS receiver devices: dedicated devices and mobile devices.

- Dedicated devices have various degrees of mobility (e.g., handheld, outdoor, or sport) suitable for hiking, bicycling, and other activities. Dedicated devices have replaceable batteries that can be run for several hours. They have small screens, some do not show color to save power, and some use transflective liquid-crystal displays to be used in bright sunlight. Their cases are sturdy and some are water resistant.
- Mobile devices are used in cars, but have a small rechargeable internal battery that can power them for an hour or two. They are installed permanently or are removable and depend entirely on the automotive electrical system.

### 1.4.2 U.S. Military/Government Model

The U.S. military/government devices similar to consumer sport products are used for commanders and regular soldiers, small vehicles, and ships. Also, the devices similar to commercial aviation products are used for aircraft and missiles. Prior to May 2000, only the United States had access to the fully accurate GPS; however, consumer devices were restricted by selective availability (SA). Therefore, the Differential GPS (DGPS) is a method to cancel the error of SA and improve GPS accuracy. The DGPS has been available in commercial applications (e.g., golf carts). The GPS is limited to about 15 meter accuracy even without SA, whereas DGPS can be within a few centimeters.

The newest precise GPS receivers are Defense Advanced Global Positioning System Receivers (DAGRs). They are smaller, accurate, and have mapping features similar to consumer GPS units.

### 1.4.3  Mapping/Resource Model

The mapping/resource devices collect location points and line area data that can be input into a Geographic Information System (GIS). They are more precise than consumer models. Their accuracy in real time can be less than a meter, and in post processing can be between 1 to 50 centimeters. They can also store more data and they are much more expensive than consumer models.

### 1.4.4  Survey Model

The survey GPS receivers are used for surveying land, where there is a need of accuracy down to the centimeter for legal or practical purposes. They are extremely precise and store a large amount of data. They are very expensive compared to consumer models and they are complex to use.

### 1.4.5  Commercial Transportation Model

The commercial transportation GPS receivers are installed on aircraft, ships, boats, trucks, and cars. They are not designed to be handheld. They provide navigation information related to the transportation. They calculate location and feed that information to large multi-input navigational computers for autopilot, course information and correction displays to the pilots, and course tracking and recording devices. They are accurate to within 10 meters. The first professional GPS receiver on the market was Trimble Navigation.

## 1.5  Application Fields of GPS Receivers

Nowadays, mobile GPS technology has enabled smartphones with convenient and highly efficient means for end users to receive navigating instructions via a global positioning system process called trilateration (Rouse 2016). Moreover, a phone's built-in GPS receiver communicates with an array of satellites, which provides navigation instructions for those either in an automobile or on foot. Therefore, advanced phones can identify individual streets and attractions on maps, as well as provide narrated tracking capability.

GPS receivers are used by civilians, pilots, boat captains, farmers, surveyors, scientists, and the military in several fields to provide accurate positioning. Among these fields, we can list aviation, marine, farming, science, surveying, military, sports, and fleet management.

### 1.5.1 Aviation

A GPS receiver is used in all aviation in all modern aircraft (AuScope 2014). It provides pilots with a real-time aircraft position and map of each flight's progress. Moreover, GPS receivers allow airline operators to pre-select the safest, fastest, and most fuel-efficient routes to each destination and ensure that each route is tracked closely when the flight is in progress.

### 1.5.2 Marine

A GPS receiver provides high accuracy for boats and ships (AuScope 2014). It allows captains to navigate through unfamiliar harbors, shipping channels, and waterways without running aground or hitting known obstacles. Moreover, a GPS receiver is used to position and map dredging operations in rivers, wharfs, and sandbars. Therefore, boat captains know precisely where it is deep enough for them to operate.

### 1.5.3 Farming

GPS receivers are used by farmers on their agricultural equipment (Andred-Sanchez and Heun 2011; AuScope 2014; Das 2013; Roberson 2005). They help farmers to map their plantations and ensure that they return to exactly the same areas when sowing their seeds in the future. In other words, the GPS receiver aims to maximize the farmers' crop production, since they rely on repeat planning season after season. It will also allow farmers to continue working in low-visibility conditions such as fog and darkness based on the GPS position's guidance instead of visual references. Moreover, a GPS receiver provides high accuracy for mapping soil locations and allows farmers to see where the soil is most fertile.

### 1.5.4 *Science*

GPS receivers are used by scientists to conduct a wide range of experiments and research, ranging from biology to physics to earth sciences (AuScope 2014). Traditionally, scientists had to tag animals with metal or plastic bands to track their various locations and monitor their movement. Nowadays, scientists can fit animals with GPS collars or tags that automatically log the animal's movement and transmit the information via satellite back to the researchers. This will provide them more detailed information about the animal's movements without having to relocate specific animals.

On the other hand, earth scientists have installed high-accuracy GPS receivers on physical features (i.e., glaciers or landslips) (AuScope 2014). This will allow them to observe and study both the speed and direction of movement, helping them to understand how landscapes change over time.

A GPS receiver can also be installed on solid bedrock to help understand very small and very slow changes in tectonic plate motion across the world (AuScope 2014).

### 1.5.5 *Surveying*

Surveyors have used GPS receivers, owing to its high accuracy, instead of line-of-sight between their instruments for mapping and measuring features on the Earth's surface and under water (AuScope 2014). They have also used GPS receivers to determine land boundaries, to monitor changes in the shape of structures, or to map the seafloor.

In the surveying domain, the GPS receiver is set up over a single point to establish a reference marker, and it is used in a moving configuration to map out the boundaries of various features as well.

### 1.5.6 *Military*

The GPS system was developed by the U.S. Department of Defense (DoD) for use by the U.S. military, but was later made available for public use (Aebi 2007; McNamara 2004). GPS navigation has been adopted by many different military forces around the world.

Nowadays, GPS receivers are used to map the location of vehicles and other assets on various battlefields in real time, in order to manage resources and protect soldiers on the ground (AuScope 2014). GPS receivers are also built in to military vehicles and other hardware such as missiles, providing them with tracking and guidance to various targets at all times of the day and in all weather conditions (Aebi 2007; AuScope 2014).

### 1.5.7 Sports

GPS receivers were applied to field sport applications in 2006 (Aughey 2011). GPS technology was rapidly used, especially in the measurement of steady-state movement. While a GPS receiver has been validated for applications for team sports, some doubts continue to exist on the appropriateness of GPS for measuring short high-velocity movements. Despite this, GPS receivers have been applied extensively in Australian football, cricket, hockey, rugby union and league, and soccer.

Moreover, GPS receivers help to collect extensive information on the activity profile of athletes from field sports, and this includes total distance covered by players and distance in velocity bands (Aughey 2011). In addition, GPS receivers have been applied to detect fatigue in matches, identify periods of most intense play, different activity profiles by position, competition level, and sport. Recently, GPS receivers have helped to integrate collected data with the physical capacity or fitness test score of athletes, game-specific tasks, or tactical or strategic information.

### 1.5.8 Fleet Management

GPS receivers are applied for fleet management in terrestrial systems (e.g., road and rail) to track the location of vehicles (Chatterjee 2009; Sharma, Kumar and Bhadana 2013; Wilson 2016). The ongoing development of GPS for autonomous vehicles also provides the greater accuracy and reliability that a vehicle requires in order to be self-driving (Novatel 2010). Therefore, without accurate GPS, accurate mapping, and an important collection of senors, they could not be considered. GPS receivers can allow real-time tracking, detect when a vehicle is outside of the predefined geofence, save fuel consumption costs by helping the driver to choose the optimal route, store up a certain number of tracking records when connection is lost, remotely power off a vehicle, control fuel consumption and temperature, detect when a vehicle is driven over the speed limit, monitor voices, and other activities.

Moreover, GPS receivers are used to secure vehicles from theft by informing the concerned users via an alert message.

### 1.6 Conclusion

We presented in this chapter a general overview of GPS and the methods of locating a position using GPS receivers to obtain accurate information. We also detailed the different segments of GPS and the different types of GPS receivers. Finally, we discussed the application fields of GPS receivers.

However, in a rapidly progressing technology world, in the next chapter we present the different electronic equipment for GPS systems and the effectiveness of each.

# 2

# ELECTRONIC EQUIPMENT FOR A GPS SYSTEM

## 2.1 Introduction

The GPS system is composed of a set of electronic equipment. The important unit is the microcontroller, which makes the GPS a smart system. The microcontroller is a small computer that contains specific applications for managing various tasks (e.g., sending alert messages, locating a given position, etc.). Microcontrollers are mainly used in products that require a degree of control to be exerted by the user (Khadse et al. 2014). Nowadays, a microcontroller is a compressed microcomputer manufactured to control the functions of many things in the technological world (e.g., embedded systems in office machines, robots, home appliances, motor vehicles, etc.). Therefore, the type of microcontroller depends on the required applications.

The GPS system is composed of other electronic components than the microcontroller, such as the GPS antenna to get the signal from the satellite that contains the geolocation of a given position whatever the weather circumstance; the camera to capture the image of a given position based on predefined characteristics or to record a video; the SD card required to store data; the audio kit to communicate via a Voice over Internet Protocol (VoIP); the cables and the connectors to link the hardware interface like the FTDI or the programmer. The FTDI or programmer communicates from the computer to the microcontroller in order to upload the required programs to the flash memory of the microcontroller.

The choice of these electronic kits depends on the client's needs.

## 2.2 Classification of Microcontrollers

The microcontrollers are classified into different categories based on the bus width, instruction set, and memory structure (Agarwal 2015).

We distinguish four basic categories of the microcontroller: the microcontroller based on the number of bits, the microcontroller based on memory devices, the microcontroller based on an instruction set, and the microcontroller based on memory architecture.

### 2.2.1  Classification of Microcontroller Based on Number of Bits

We distinguish three types of microcontrollers based on the number of bits (Agarwal 2015; Kamal 2012; Parai et al. 2013): an 8-bit microcontroller, a 16-bit microcontroller, and a 32-bit microcontroller.

- The 8-bit microcontroller has an internal bus of 8-bit (e.g., Intel 8031/8051) where the Central Processing Unit (CPU) or the Arithmetic Logic Unit (ALU) can process 8-bit data. This type of microcontroller is used in position control and speed control.
- The 16-bit microcontroller performs greater precision and good performance as compared to the 8-bit microcontroller. This type of microcontroller is used in high-speed applications such as servo control systems, robotics, and other applications. For example, extended Intel 8096 and Motorola MC68HC12 families are considered to be 16-bit microcontroller units.
- The 32-bit microcontroller uses the 32-bit instructions to perform the arithmetic and logic operations. This type is developed for the purpose of very high-speed applications in image processing, telecommunications, intelligent control systems, and other applications. For example, the Intel/Atmel 251 family, the PIC3x, and the ARM are considered to be 32-bit microcontroller units.

### 2.2.2  Classification of Microcontroller Based on Memory Devices

We distinguish two types of microcontrollers based on memory devices (Agarwal 2015; Kamal 2012; Mazidi et al. 2013): embedded memory microcontrollers and external memory microcontrollers.

- Embedded memory microcontrollers are embedded systems that have a microcontroller unit where all the functional blocks are available on a chip. For example, the 8051 that has program and data memory, Input/Output (I/O) ports, serial communication, counters and timers, and interrupts on the chip is considered to be an embedded microcontroller.

- External memory microcontrollers are embedded systems that have a microcontroller unit where not all the functional blocks are available on a chip. For example, the 8031 has no program memory on the chip and is an external memory microcontroller.

### 2.2.3 *Classification of Microcontroller Based on the Instruction Set*

We distinguish two types of microcontrollers based on the instruction set (Agarwal 2015; Kamal 2012): a complex instruction set computer (CISC) and a reduced instruction set computer (RISC).

- The complex instruction set computer (CISC) allows the user to apply one instruction as an alternative to many simple instructions.
- The reduced instruction set computer (RISC) reduces the operation time by shortening the clock cycle per instruction. The reduced instruction set computer has better execution than the complex instruction set computer.

### 2.2.4 *Classification of Microcontroller Based on Memory Architecture*

We distinguish two types of microcontrollers based on memory architecture (Agarwal 2015; Kamal 2012): Harvard memory architecture microcontrollers and Princeton memory architecture microcontrollers.

- The Harvard memory architecture microcontroller unit has a Harvard memory architecture in the processor when it has a dissimilar memory address space for the program memory and the data memory.
- The Princeton memory architecture microcontroller has the microcontroller unit with Princeton memory architecture in the processor when it has a common memory address for the program memory and the data memory.

## 2.3 Types of Microcontrollers

We distinguish four types of microcontrollers: the 8051 microcontroller, the Peripheral Interface Controller (PIC) microcontroller, the RISC Machines (ARM) microcontrollers, and the Alf-Egil Bogen and Vegard Wollan's RISC (AVR) microcontrollers.

### 2.3.1 8051 Microcontroller

The 8051 microcontroller was invented by the Intel Corporation in 1981 (Agarwal 2015, 2016; RoseMary 2010). It is an 8-bit microcontroller and is available in 40 pin dual in line (DIP). It is the basic microcontroller and still manufactured by many companies because of its facility for allowing its integration into a device. The recent 8051 microcontroller has six clock cycles per instruction. The 8051 microcontroller is considered to be a CISC processor and does not have an inbuilt memory bus and Analog/Digital (A/D) converters. The 8051 microcontrollers are applied to a wide range of devices (e.g., energy management, touch screens, automobiles, and medical devices) (Agarwal 2015, 2016).

The 8051 microcontroller is designed to have a strict Harvard architecture (Rodriguez 1995). It can only execute code fetched from program memory and has no instructions to write to program memory. However, some 8051 microcontrollers have some dual-mapped RAM making them act like they have a Von Neumann architecture; since the external ROM and RAM share data and address buses, then the mapping can be designed to allow Read/Write data access to program memory.

The features of an 8051 microcontroller are the following (Agarwal 2015):

- 8-bit Central Processing Unit (CPU)
- 16-bit Program Counter
- 8-bit Processor Status Word (PSW)
- 8-bit Stack Pointer
- 4 kbytes internal Read Only Memory (ROM) (i.e., program memory)
- 128 bytes internal Read Access Memory (RAM) (i.e., data memory)

- Special Function Registers (SFRs) of 128 bytes
- 32 Input/Output (I/O) pins arranged as four 8-bit ports (i.e., P0, P1, P2, and P3)
- Two 16-bit timer/counters (i.e., T0 and T1)
- Two external and three internal vectored interrupts
- One full duplex serial I/O

### 2.3.2 Peripheral Interface Controller (PIC) Microcontroller

The Peripheral Interface Controller (PIC) Microcontroller, with Harvard architecture, is a family of microcontrollers by Microchip Technology USA (Agarwal 2015, 2016; Mazidi et al. 2013). In 1993 it was developed by General Instrument's Microelectronics as a supporting device for program data processor (PDP) computers to support its peripheral devices. The PIC microcontroller is controlled by the software and could be programmed to complete many tasks, control a generation line, and other functions. The PIC microcontroller is a RISC processor. The PIC microcontroller has a machine cycle with four clock pulses in contrast to 12 clock pulses in an Intel 8051 microcontroller.

The PIC microcontroller is applied in smart phones, audio accessories, video gaming peripherals, and advanced medical devices (Agarwal 2015).

At the start, Microchip manufactured PIC16F84 and PIC16C84, which were the only affordable flash PICs (Agarwal 2016). Recently, Microchip has introduced flash chips with more attractive types (e.g., 16F628, 16F877, and 18F452). The 16F877, as compared with the PIC16F84, is more expensive and has more features (e.g., code size, RAM, I/O pins, A/D converter, etc.).

The features of the PIC 16F877 microcontroller are the following (Agarwal 2016):

- High-performance RISC CPU
- Up to 8K × 14 words of FLASH program memory
- 35 Instructions (fixed length encoding-14-bit)
- 368 × 8 static RAM-based data memory
- Up to 256 × 8 bytes of Electrically Erasable Programmable Read-Only Memory (EEPROM) data memory
- Interrupt capability (up to 14 sources)

- Three addressing modes (direct, indirect, relative)
- Power-on reset (POR)
- Harvard architecture memory
- Power saving SLEEP mode
- Wide operating voltage range: 2.0 V to 5.5 V
- High sink/source current: 25 mA
- Accumulator-based machine

### 2.3.3 Advanced RISC Machines (ARM) Microcontroller

Advanced RISC Machines (ARM) microcontrollers are a family of microcontroller based on RISC architecture and developed by ARM Limited (Agarwal 2016; Electronics Hub 2017). ARM microcontroller makes 32-bit and 64-bit RISC multi-core microcontrollers. An ARM microcontroller has von Neumann architecture (i.e., the program and the RAM in the same space). ARM Microcontrollers are much in use for power saving and operate in very low power consumption. They can operate at a higher speed and perform extra millions of instructions per second (MIPS), since they are designed to perform a smaller number of types of computer instructions. ARM microcontrollers are widely used in modern handsets for mobile communications. Also, they are used in various other electronic embedded systems (e.g., handhelds, smart phones, tablets, multimedia players, disk drivers, etc.).

The features of the ARM microcontroller are the following (Agarwal 2016; Electronics Hub 2017):

- Maximum single cycle functioning
- Constant 16 × 32 bit register file
- Load or store architecture
- Preset instruction width of 32 bits so as to simplify pipe-lining and decoding, at minimized code density

### 2.3.4 Alf-Egil Bogen and Vegard Wollan's RISC (AVR) Microcontroller

The Alf-Egil Bogen and Vegard Wollan's RISC (AVR) micro-controller is a modified Harvard RISC architecture 8-bit RISC single-chip microcontroller, which in 1996 was developed by Atmel Corporation (Agarwal 2015, 2016). The AVR takes only one clock per instruction. The AVR microcontroller has separate memories for data and programs. AVRs operate at high speed compared to 8051 and PIC microcontrollers.

AVR microcontrollers are classified into three types (Agarwal 2015, 2016): TinyAVR, MegaAVR, and XmegaAVR.

- The TinyAVR has less memory, a small size, and is suitable only for simpler applications.
- The MegaAVR is the most popular microcontroller having a good amount of memory (up to 256 KB), a higher number of inbuilt peripherals, and is suitable for moderate to complex applications.
- The XmegaAVR is used commercially for complex applications, which require large program memory and high speed.

The features of the AVR microcontroller are the following (Agarwal 2015):

- 16KB of In-System Programmable Flash
- 512B of In-System Programmable EEPROM
- 16-bit Timer with extra features
- Multiple internal oscillators
- Internal, self-programmable instruction flash memory up to 256K
- In-system programming using ISP, JTAG, or high-voltage methods
- Optional boot code section with independent lock bits for protection
- Synchronous/asynchronous serial peripherals (UART/USART)
- Serial peripheral interface bus (SPI)
- Universal serial interface (USI) for two/three-wire synchronous data transfer
- Watchdog timer (WDT)
- Multiple power-saving sleep modes
- 10-bit A/D converters, with multiplex of up to 16 channels
- CAN and USB controller support
- Low-voltage devices operating down to 1.8 v

There are many AVR family microcontrollers, such as the ATmega8, the ATmega16, the ATmega328, and others. The ATmega328 has a flash memory of 32 kB, whereas the ATmega8 has 8 kB of flash memory.

Some features of AVR microcontrollers are given as follows (Agarwal 2015):

- 28-pin AVR microcontroller
- Flash program memory of 32 kbytes
- EEPROM data memory of 1 kbyte
- SRAM data memory of 2 kbytes
- I/O pins are 23
- Two 8-bit timers
- A/D converter
- Six-channel pulse-width modulator (PWM)
- Inbuilt USART
- External Oscillator: up to 20 MHz

### 2.4 Comparison of the Different Microcontroller Types

Table 2.1 represents a comparative study among 8051, PIC, ARM and AVR microcontroller types, as well as the advantages and the disadvantages of each type (Agarwal 2015, 2016; Electronics Hub 2017; Ozden 2013). The 8051 and PIC microcontrollers need multiple clock cycles per instruction, whereas ARM and AVR microcontrollers execute most instructions in a single clock cycle. Therefore, the speed of ARM and AVR microcontrollers is more than the speed of 8051 and PIC microcontrollers, with the PIC microcontroller being faster than the 8051 microcontroller. The 8051 microcontroller is an 8-bit microcontroller based on CISC architecture, but the AVR microcontroller is an 8-bit microcontroller based on RISC architecture. The 8051 microcontroller consumes more power than ARM and AVR microcontrollers. We can program more easily in an 8051 microcontroller than in an AVR microcontroller. ARM microcontrollers have smaller size, reduced difficulty, and lower power expenditure, which make them suitable for increasingly miniaturized devices.

**Table 2.1**  Comparison of the Different Microcontroller Types

| | 8051 MICROCONTROLLER | PIC MICROCONTROLLER | ARM MICROCONTROLLER | AVR MICROCONTROLLER |
|---|---|---|---|---|
| Bus Width | 8 bit | 8 bit, 16 bit, and 32 bit | 32 bit mostly and 64 bit | 8 bit and 32 bit |
| Speed | 12 clocks per instruction cycle | 4 clocks per instruction cycle | 1 clock per instruction cycle | 1 clock per instruction cycle |
| Memory | ROM, SRAM, and Flash | SRAM and FLASH | SDRAM, FLASH, and EEPROM | SRAM, FLASH, and EEPROM |
| Instruction Set Architecture | CLSC | Some features of RISC | RISC | RISC |
| Memory Architecture | Harvard architecture and some Von Neumann architecture | Harvard architecture | Von Neumann architecture | Modified Harvard architecture |
| Power Consumption | Average | Low | Low | Low |
| Families | 8051 variants | PIC16, PIC17, PIC18, PIC24, and PIC32 | ARMv1, 2, 3, 4, 5, 6, 7, and 8 series | TinyAVR, AtmegaAVR, XmegaAVR |
| Cost | Very low | Average | Low | Average |
| Other Features | Known for its standard | Cheap | High-speed operation and vast | Cheap and efficient |

## 2.5 Future Technology Devices International (FTDI)

The Future Technology Devices International (FTDI) chipset is a hardware peripheral converter from a RS232 serial data or TTL serial transmissions (i.e., sending data one bit at a time) microcontroller side into a Universal Serial Bus (USB) computer side, in order to allow support for legacy devices with modern computers (FTDI Chip 2017). In other words, the FTDI, by creating a virtual serial port, establishes a serial communication between the computer and the microcontroller. This technology is considered a better performing and faster method to program the microcontroller than the programmer (cf. Section 2.6).

## 2.6 Programmer

A microcontroller programmer or microcontroller burner is a hardware device accompanied by software, which is used to transfer the machine language code to the flash memory of the microcontroller from the computer (AllAboutEE 2012; Atmel 2016; Choudhary 2012). The compiler converts the code written in programing language (e.g., assembly, C, Java, etc.) to machine language code (i.e., understandable by the microcontroller) and stores it in a hexadecimal file. In other words, the microcontroller programmer acts as an interface between the computer and the microcontroller. The application software interface (API) of the programmer reads data from the hexadecimal file stored on the computer and supplies it into the flash memory of the microcontroller. Therefore, to burn (i.e., program) a microcontroller, we should take it out of the circuit, place it on a burner (i.e., programmer) and then dump the hexadecimal file into the microcontroller using the API. Nowadays, the latest microcontrollers have the feature (i.e., bootloader memory), which allows self-burning capabilities without the need of additional programmer hardware. They need only an API to transfer the program to the microcontroller. Therefore, this API can be incorporated in the compiler, which can directly burn the microcontroller.

## 2.7 Bootloader

According to Electrical Engineering (Electrical Engineering 2012), the bootloader is described as a program that runs in the microcontroller to be programmed. The bootloader is defined by EngineersGarage (EngineersGarage 2012) as a small section in the ROM of the microcontroller, which executes first when it is initialized. In other words, it is the first program which executes, before the main program, whenever a system is initialized. The bootloader always runs from reset (Peatman 1988).

The bootloader enriches the capabilities of the microcontroller and makes it a self-programmable device (Electrical Engineering 2012; EngineersGarage 2012). Therefore, the bootloader receives new program information externally via some communication means (i.e., FTDI chipset [cf. Section 2.5] or programmer [cf. Section 2.6]) and writes that information to the flash memory (i.e., program memory) of the microcontroller. As the new code arrives, the old code is overwritten. A checksum (NPTEL 2009) is included with the uploaded code, thus the bootloader can tell if the new application is intact. In the case of the new application not being intact, it will stay in the bootloader constantly requesting an upload until receiving into memory one with a valid checksum. On the other hand, a user runs a program on the computer, which communicates to the bootloader existing in the microcontroller. This sends the new binary to the bootloader, which writes it to flash memory, and then generates the new code to be run.

## 2.8 Cables and Connectors

The cables and the connectors together serve to communicate via the different elements of the GPS system in order to exchange data (e.g., programs, signal information, requests, etc.). Among the cables and connectors, we can list the Transmit (TX), the Receive (RX) (Pololu 2015), and the Universal Serial Bus (USB). TX aims to send requests (e.g., send e-mail, send position, etc.) from the microcontroller to the Global System for Mobile Communication (GSM) system network (i.e., 3G network or radio frequency) (GSMA 2017) as well as to transmit information from the microcontroller as serial bytes (i.e., sending data one bit at a time) to be visualized on the screen of the Integrated Development Interface (IDE) (Silverthorne 2016) installed on the connected computer through the Future Technology Devices International (FTDI) chipset (FTDI Chip 2017). The TX also sends a request for Parity bit (i.e., check bit for error detection) (Swarthmore 2010) to the connected computer through the FTDI chipset or the programmer (Atmel 2016). The FTDI chipset and the programmer are connected to the computer via a USB port. This USB port has two objectives. The first is to power the microcontroller. The second is to create a link between the computer and the microcontroller through the FTDI chipset or the programmer in order to upload the developed programs of type hexadecimal in the flash memory of the microcontroller. The FTDI chipset is connected to the microcontroller via a serial (i.e., a Universal Asynchronous Receiver/Transmitter [UART]) connector composed of six pins (i.e., RX, TX, Power supply negative or Ground [GND], and Supply voltage + [VDD or VCC], with a Reset to reset the microcontroller with software, and the Clock [CLK], where the frequency of operation of the microcontroller is determined by the clock cycle) (FTDI Chip 2017; Peatman 1988). The programmer is connected to the microcontroller via a serial peripheral interface (SPI) bus composed of six pins (i.e., Master Output Slave Input [MOSI], Master Input Slave Output [MISO], Serial Clock [SCLK], Slave Select [SS] or Chip Select [CS] as reset button, and Serial Data I/O [SDIO] in bidirectional, Ground [GND]) (Peatman 1988; Swarthmore 2010). The RX helps to receive the information from the GPS module and the GSM system network, as well as to receive the hexadecimal developed programs from the IDE via the FTDI or the programmer (AllAboutEE 2012; Choudhary 2012).

## 2.9 SD Card

The SD Card is a hardware component built in the GPS module. It can store map files, geocaches, routes, waypoints or track files, custom points of interest (POIs), with all the data coming from the 3G network, the photos and video files recorded by the GPS camera, the video call files reproduced during a video call, or the audio files (Garmin 2011; Surratt 2003). Moreover, the SD card could be a reader/writer, thus we can upload a file from the SD card to File Transfer Protocol (FTP) server and vice versa, as well as can be connected to a computer using the USB port. Using a SD writer, the upload rarely takes more than 10 minutes, even for the largest cards. In addition, with the SD card socket, we can handle a complete FAT32 file system and store up to 32 GB of information (Blacksys 2015). This allows the 3G module to work at full speed without the need of other alternative storing devices. This is because transferring of large data over a serial port can take hours even for the smallest SD cards (Surratt 2003).

## 2.10 GPS Antenna

A GPS receiver needs to receive a signal from as many satellites as possible (U-blox 2009). However, GPS signals are extremely weak and present unique demands on the antenna for GPS performance. We distinguish two types of GPS antenna: an internal GPS antenna and an external GPS antenna. Both of the antennas are used with a GPS module to communicate with the satellites in order to get a geolocation (i.e., longitude, latitude, and time) via a signal for a given position at a given time. The GPS antenna will boost the performance of a GPS receiver, especially in good sky visibility. On the other hand, the optimal performance of the GPS will not be available in obstacle locations (e.g., narrow streets, underground parking, objects covering the antenna, etc.).

The GPS antenna requirements for an optimal GPS performance are the following (Motschenbacher and Connelly 1993; U-blox 2009):

- An active antenna, which contains active built-in electronic components (e.g., transistors) in order to have a wider frequency range (i.e., bandwidth) and helps to keep the receiver noise figure low.
- A low level of directivity, that measures the degree to which the radiation emitted is concentrated in a single direction.
- Good antenna visibility for the sky.
- Good matching between antenna and cable impedance. This represents the ratio of the amplitudes of voltage and current of a single wave propagating along the line and travelling in one direction when there is an absence of reflections in the other direction.
- High gain (e.g., >4 dBi), which represents the ratio of the power produced by the antenna to the power produced by a hypothetical lossless isotropic antenna. It increases with the

level of directivity. The gain unit is represented in decibel iso-
tropics (dBi).

- A low-noise amplifier (LNA) (e.g., <2 decibels [dB] noise
  figure), which amplifies a very low-power signal and mini-
  mizes additional noise.
- A filter, to perform signal processing function by removing
  the unwanted frequency components from the signal and
  enhancing the wanted ones.

### 2.11 GPS Camera

A GPS camera aims to take photos and to pinpoint the exact location at which the image is taken, as well as to record video in high
resolution (i.e., 640 × 480 video graphics array [VGA], 1600 × 1200
ultra eXtended graphics array [UXGA]) for a given location at a given
time; and then to send them to a predefined destination via various communication protocols (i.e., HTTP, HTTPS, FTP, FTPS,
SMTP, and POP3) that we will detail in Chapter 3 (Génération
Robots 2017). Moreover, the data transferred to the computer including the geographic location allows tagging this location on a map
(CameraDecision 2016) or to remember later where this photo had
been taken (Schurman 2017). This camera is connected directly to
the GPS module, which enables the record of videos in high resolution and to do a photo based on a scheduled task or a received command from a predefined user. Moreover, a GPS camera is used when
performing a video call. The GPS camera is configured with some
parameters (e.g., resolution, frames per second [FPS], rotation, or
zoom) (Génération Robots 2017).

### 2.12 Audio Kit

An audio kit includes microphone, speaker, hands free and headphone sets connected to the GPS module using pin connectors. A
specific input and output of the selected audio kit should be activated
on the GPS module. An audio kit, if a camera is involved, allows
recording and playing sound, as well as to receive voice calls or video
calls. While out surveying, an audio kit is used in audio mapping
(i.e., voice mapping) to record data (i.e., street names, street types,

access restrictions, points of interests [POIs]) (OpenStreetMap 2016). Moreover, GPS systems with an audio kit offer recording of conversations, including confidential ones, without the knowledge of the people involved (Christopher 2014).

### 2.13 Conclusion

In this chapter we introduced the classification of microcontrollers based on the number of bits, memory devices, instruction sets, and memory architecture. Thus, we have presented various types of microcontrollers with their effectiveness. This study helped us to choose a microcontroller providing efficient control quickly and with less cost. Future Technology Devices International (FTDI) and the programmer are shown to present the converter means from serial to USB in order to send the developed programs in the Integrated Development Environment (IDE) computer to the microcontroller (i.e., bootloader in case of initiation program and flash memory), and to send the information for display in the reverse direction. Similarly, cables and connectors, SD card, GPS antenna, GPS camera and audio kits are detailed to present the additional electronic equipment for a GPS system.

An interest in the study of electronic equipment for a GPS system is used to justify our choice. To present the means of communication among the different components of the GPS system, we integrate the various communication modules and the different communication protocols in the next chapter.

# 3

# COMMUNICATION MODULES
# AND PROTOCOLS

## 3.1 Introduction

The GPS system is composed of a set of communication modules (e.g., Global Positioning System [GPS] module, Global System for Mobile Communications [GSM] module, satellite module, etc.) and a set of electronic equipment (e.g., FTDI, programmer, etc.). These modules and equipment interact together via specific communication protocols (i.e., microcontroller communication protocols and Internet communication protocols) to achieve a given task.

The GPS module is a device that uses the Global Positioning System to determine the location of a given object (e.g., vehicle, person, etc.) at anytime and anywhere on the Earth (Electronics Hub 2015).

The GSM module is a digital circuit-switched network for data communication (e.g., by e-mail, via text messages SMS, etc.) (Sheriff and Fun Hu 2001).

The satellite module is a specialized wireless receiver/transmitter placed in orbit around the Earth to relay and amplify radio telecommunications signals simultaneously via a transponder (Huurdeman 2003).

The microcontroller communication protocols aim for communication by the microcontroller with external devices (e.g., FTDI, SPI, etc.) or communication with the different communication modules (e.g., GPS module, GSM module, etc.).

Other than the microcontroller communication protocols, we can mention the Internet communication protocols (Alani 2014; Comer 2000; Edwards and Bramante 2009; Forouzan 2000). These last protocols use the Transport Communication Protocol (TCL) to provide different application services (e.g., Hypertext Transfer Protocol

[HTTP] for Web navigation, Simple Mail Transport Protocol [SMTP] for the transmission of the e-mails, File Transfer Protocol [FTP] for the transfer of files, etc.)

The choice of these communication modules and these communication protocols depends on the requirements of the GPS system's application.

## 3.2 Different Communication Modules

We distinguish three different communication modules: the Global Positioning System (GPS) module, the Global System for Mobile Communications (GSM) module, and the Satellite module.

### 3.2.1 Global Positioning System (GPS) Module

A GPS module is a device that uses the Global Positioning System to determine the location of a vehicle or person (Electronics Hub 2015). GPS receivers are used to provide reliable navigation, positioning, and timing services to the users at anytime and anywhere on the Earth (cf. Chapter 1, Section 1.1).

Each satellite transmits the messages continuously, which contains the time they were sent. The GPS module calculates the position by reading the signals that are transmitted by the satellites (Electronics Hub 2015). In other words, the information received from the signal of four satellites (i.e., three GPS satellites triangulate and measure the distance to the GPS receiver based on the arrival time of each message and compare the measurements; and a fourth satellite measures the time to the GPS receiver) is compiled to determine the location of the GPS receiver (Brown, McCabe and Welford 2007). The received raw data is converted for the user to latitude, longitude, altitude, speed, and time. Moreover, GPS receivers, capable of receiving Differential GPS (DGPS) signals and using that information in computing a location, can theoretically derive location accuracy to about 10 cm (as opposed to nominal accuracy of about 5 m when using only traditional GPS signals) (Brown, McCabe and Welford 2007).

Therefore, GPS receivers are basically radio receivers equipped to receive the GPS satellite transmissions (Brown, McCabe and Welford 2007). However, it is necessary to equip the GPS receivers with "channels" in order to be able to receive the satellites' frequencies. Furthermore, it is necessary to have in a GPS receiver a minimum of four channels, since GPS receivers compute location based on signals from at least four simultaneous satellite signals.

The GPS module continuously transmits the received data as per the NMEA standards (Electronics Hub 2015) to the microcontroller serially using the UART protocol (cf. Section 3.4.3.1) or via SPI protocol (cf. Section 3.4.3.2). The microcontroller extracts the latitude and longitude values from the received messages and displays the position on a display screen (i.e., liquid crystal display [LCD]) that shows a map, or sends them by e-mail or via text message (Rouse 2016).

There are three basic concepts involved in transmission of the collected GPS location data to a vendor's software for processing (Brown et al. 2007): active, passive, and hybrid.

- *Active:* The communication is accomplished using cellular communication technology and in a near real-time basis (e.g., every minute, every five minutes, etc.). The time parameter is configurable based on a specific client.
- *Passive:* The GPS data is collected by the GPS receiver throughout the day. The GPS receiver is connected to a charging unit to transmit the collected data points from the day for processing using a landline phone connection.
- *Hybrid:* The collected data is sent to the vendor using cellular communication technology on a less regular basis (i.e., every few hours), which is programmable, but automatically switches to an active mode in the event of an alert. For a Hybrid, the time parameter is usually much longer than with an Active concept, but more frequently than once a day as with a Passive concept.

The vendor's software is accessible via the Internet, which allows agencies to access their client GPS data from any computer connected to the Internet using a Web browser.

Moreover, some clients use radio frequency (RF) technology to verify the proximity between a tamper-resistant bracelet and a stationary device, where the bracelet will transmit an alert via the GPS receiver if tampering occurs (i.e., when the bracelet and the stationary device exceed the established distance parameters [e.g., 100 feet] during predetermined time frames) (Brown et al. 2007). Instead of using a stationary device, most vendors utilize extended RF technology for GPS solutions by configuring the RF signal to communicate between the tamper-resistant bracelet and the GPS receiver. In addition, some vendors have eliminated the need for RF by designing a single device that is both a GPS unit and a tamper-resistant bracelet.

### 3.2.2 Global System for Mobile Communications (GSM) Module

By the middle of the 1980s, the mobile industry's attention had focused on the need to implement more spectrally efficient 2G digital type services, offering a number of significant advantages including greater immunity to interference, increased security, and the possibility of providing a wider range of services (Sheriff and Fun Hu 2001). The 2G networks were developed as a replacement for first generation (1G) analog cellular networks, and the Global System for Mobile Communications (GSM) standard that was originally described as a digital circuit-switched network optimized for full duplex voice telephony. This expanded over time to include data communications, first by circuit-switched transport, then by packet data transport via general packet radio service (GPRS) and Enhanced Data rates for GSM Evolution (EDGE) or EGPRS. The Global System for Mobile Communications (GSM), originally the Groupe Spécial Mobile, is a standard employing time-division multiple-access (TDMA) spectrum-sharing developed by the European Telecommunications Standards Institute (ETSI). It was used for the protocols for second-generation (2G) digital cellular networks used by mobile phones. It was first deployed in Finland in December 1991 (Huurdeman 2003; Sauter 2014; Sheriff and Fun Hu 2001). In 1992, the sending and the receiving of the first short messaging service (i.e., SMS or "text message" up to 160 characters in length) was completed, and Vodafone UK and Telecom Finland signed the first international roaming agreement. In 1995, fax, data, and SMS messaging services were launched commercially. At the end of 1998, ETSI completed its standardization of the GSM Phase 21 services high speed circuit switched data (HSCSD) and GPRS. These two new services aimed to exploit the potential markets in the mobile data sector and recognize the influence of the Internet on mobile technologies. The GSM standard does not include the 3G Universal Mobile Telecommunications System (UMTS) code division multiple access (CDMA) technology nor the 4G LTE orthogonal frequency-division multiple access (OFDMA) technology standards issued by the 3GPP. In 2000 the first commercial GPRS services were launched, and the first GPRS-compatible handsets became available for sale. In 2001 the first UMTS (W-CDMA) network was launched. This is a 3G technology that is not part of GSM. In 2003, EDGE services became operational in a network. Subsequently, the 3GPP developed third-generation (3G) UMTS standards followed by fourth-generation (4G) LTE Advanced standards, which do not form part of the ETSI GSM standard.

GSM employs a number of logical control channels to manage its network (Sheriff and Fun Hu 2001). These channels are grouped under three categories: a broadcast control channel (BCCH), a common control channel (CCCH), and a dedicated control channel (DCCH).

Telstra in Australia shut down its 2G GSM network on December 1, 2016, then AT&T Mobility from the United States shut down its GSM network on January 1, 2017, and Singapore phased out 2G services by April 2017 (AT&T 2016; Singtel 2015; TeleGeography 2016).

### 3.2.3 Satellite Module

According to Rouse (Rouse 2008, 2016), a satellite is an artificial object which has been intentionally placed into orbit. In a communication context, it is a specialized wireless receiver/transmitter that is launched by a rocket and placed in orbit around the Earth (Huurdeman 2003). It relays and amplifies radio telecommunications signals simultaneously via a transponder (Sheriff and Fun Hu 2001). It creates a communication channel between a source transmitter and a receiver at different locations on Earth. On the other hand, the satellites aim to relay the signal around the curve of the Earth allowing communication between widely separated points. Moreover, the Differential GPS (DGPS) technique improves the accuracy of conventional satellite positioning by transmitting a "correction signal" from a fixed ground station that represents the difference between the GPS-derived location of the station and the known location of the station (Differential GPS) (Brown et al. 2007).

GPS satellite transmissions are essentially radio waves transmitted on specific frequencies (Brown et al. 2007). However, GPS satellites transmit their signals on several different channels representing very narrow frequency bands. In general, the band is related to the age of the satellite. Older satellites in the constellation use what is known as the L1 frequency. Newer satellites also broadcast on frequencies known as L2 and L2C. Finally, planned future satellites will add a new frequency called L5 (Satellite Navigation: GPS Modernization).

The communications satellites are classified into two categories (Aerospace 2010): passive satellites and active satellites.

- Passive satellites only reflect the signal coming from the source toward the direction of the receiver, where the signal received from Earth is very weak due to free-space path loss (i.e., the satellite is far above the Earth).
- Active satellites amplify the received signal before retransmitting it to the receiver on the ground.

There are three types of communications satellites categorized according to the type of orbit they follow (GCMD 2008; Rouse 2008; Sheriff and Fun Hu 2001): geostationary orbit (GEO), medium Earth orbit (MEO), and low Earth orbit (LEO).

- Geostationary satellites have a geostationary orbit (GEO), which is 35,786 kilometers (22,236 miles) from the Earth's surface. The advantage of this orbit is that ground antennas do not have to track the satellite across the sky; they can be fixed to a point at the location in the sky where the satellite appears.
- Medium Earth orbit (MEO) satellites are closer to the Earth. Orbital altitudes range from 2,000 to 35,786 kilometers (1,243 to 22,236 miles) above the Earth.
- Low Earth orbit (LEO) is the region below medium orbits. It is about 160 to 2,000 kilometers (99 to 1,243 miles) above the Earth.

As satellites in MEO and LEO orbit the Earth faster, they do not remain visible in the sky relative to a fixed point on Earth continually like a geostationary satellite, but appear to a ground observer to cross the sky and "set" when they go behind the Earth (GCMD 2008;

Rouse 2008). Therefore, to provide continuous communication capability with these lower orbits requires a larger number of satellites, so one satellite will always be in the sky for transmission of communication signals. However, due to their relatively small distance from the Earth, their signals are stronger.

There are three main systems of satellite networks (Advanced Tracking 2012): Inmarsat, Globalstar, and Iridium. Inmarsat and Iridium offer global coverage, whereas Globalstar offers a smaller coverage, but is still very important. Also, only the Inmarsat network provides real-time tracking.

As per Electronic Hubs (Electronics Hub 2015), the Global Positioning System (GPS) uses 24 to 32 satellites to provide the data to the receivers. Satellites are used for diverse purposes and our focus is on the Global Positioning System (GPS).

According to Rouse (Rouse 2016), the GPS satellites are a constellation of approximately 30 well-spaced satellites that orbit the Earth and make it possible for people with ground GPS receivers to pinpoint their geographic location. The location accuracy is anywhere from 100 to 10 meters for most equipment. There are 21 GPS satellites and three spare satellites that orbit at 10,600 miles above the Earth. The satellites are spaced, so that from any point on Earth, four satellites will be above the horizon. Each satellite contains a computer, an atomic clock, and a radio.

### 3.3 Recapitulative Table of Comparison of Different Communication Modules

Table 3.1 represents a comparative study between different communication modules by showing the differences, as well as the accuracy and the usage limit of each (i.e., GPS, GPRS, GSM, and Satellite) (Advanced Tracking 2012; Difference Between 2017; Esa 2013; LiveViewGPS 2007; RF Wireless World 2012).

### 3.4 Different Communication Protocols

Among the different protocols for communication of microcontrollers, we can mention the Universal Asynchronous Receiver/Transmitter (UART) protocol and the Serial Peripheral Interface (SPI) protocol.

**Table 3.1** Comparison of Different Communication Modules

| SPECIFICATIONS | GPS | GPRS | GSM | SATELLITE |
|---|---|---|---|---|
| Service | Positioning service | Data service used in mobile phones | Mobile communications using SIM card | Satellite networks |
| Technologies | Wireless (e.g., GSM, LTE, etc.) | GSM | Variation of time division multiple access (TDMA) | Radio frequency (RF) |
| Communication with tiers | Collection of satellites that orbit Earth Terrestrial cellular towers | Terrestrial tower | One cellular tower | GPS receivers |
| Functionalities | Receives satellite signal to determine its precise location Identify geolocation (i.e., latitude, longitude, time, speed, etc.) for various applications on land, sea, and air Triangulation to at least three or four of the 24 satellites that orbit the Earth | Exchange of data Used for reception applications (e.g., Internet, e-mail, SMS, MMS, etc.) | Communicates data over the service provider's cellular network An object's position is determined using signal strength and triangulation from base stations | Relay television and radio signals from the broadcast point to stations around the country |
| Frequency of operation | Works on 1575.42 MHz (L1 Band), 1227.60 MHz (L2 Band), 1381.05 (L3 Band), 1379.913 (L4 Band), 1176.45 (L5 Band), etc. | Works on various GSM bands (e.g., 800 MHz, 900 MHz, 1800 MHz, 1900 MHz, etc.) | The GSM expertise uses five bands at MHz rate (i.e., 450 MHz, 850 MHz, 900 MHz, 1800 MH, and 1900 MHz) | Works on L-band (1–2 GHz), S-band (2–4 GHz), C-band (4–8 GHz), X-band (8–12 GHz), Ku-band (12–18 GHz), and Ka-band (26–40 GHz) |

(Continued)

**Table 3.1 (*Continued*)**  Comparison of Different Communication Modules

| SPECIFICATIONS | GPS | GPRS | GSM | SATELLITE |
|---|---|---|---|---|
| Requirements to work | At least three satellites, due to basic principles of trilateration, or four satellites to determine precise position | Just one station | At least four satellites to determine precise position | GSM network |
| Accuracy | Accuracy within a meter<br>Works in any kind of weather condition<br>Helps in searching the local area for nearby amenities | | Accuracy within 10 meters<br>Base stations are capable of providing locations in areas like tunnels and dense areas | Located everywhere (e.g., sea, desert, city, etc.) |
| Usage limit | Anywhere when sky is visible<br>Comparatively difficult in area surrounded by tall buildings | Limited in range to cellular towers located on land | Worldwide roaming with expensive charges<br>Loss of connection when outside of coverage | No roaming charges<br>Real security when outside of coverage |

### 3.4.1 Communication Protocols of Microcontrollers

Communication between electronic devices is like communication between human beings. Both sides need to speak the same language, which is called a communication protocol as an electronic concept. Among the different protocols of communication, we can mention the Universal Asynchronous Receiver/Transmitter (UART) protocol, the Serial Peripheral Interface (SPI) protocol, and the Inter-Integrated Circuit (I2C) protocol (Circuit Basics 2016). These protocols are quite a bit slower than protocols like the USB, Ethernet, Bluetooth, and WiFi. However, they use less hardware and system resources; and they are ideal for communication between microcontrollers and between microcontrollers and sensors where there is no need to transfer large amounts of high-speed data.

*3.4.1.1 Universal Asynchronous Receiver/Transmitter (UART) Protocol* The Universal Asynchronous Receiver/Transmitter (UART) is a communication protocol based on a physical circuit in a microcontroller or a stand-alone integrated circuit (IC) (Circuit Basics 2016). Two UARTs communicate directly to each other. The UART converts parallel data (i.e., all bits of data are sent at the same time, each through a separate wire) from a controlling device (i.e., central processing unit [CPU]) into serial form (i.e., one bit at a time), transmits it in serial to the receiving UART, which then converts the serial data back into parallel data for the receiving device. The UART uses two wires to transmit data between the UART's devices. The data flows are done via the Tx pin of the transmitting UART device to the Rx pin of the receiving UART device, as shown in Figure 3.1.

UARTs transmit data asynchronously. In other words, there is no signal clock to synchronize the output of bits from the transmitting

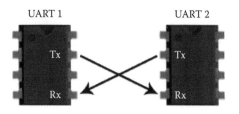

**Figure 3.1** UART basic connection diagram. (Source: Circuit Basics. 2016. Basics of UART Communication. http://www.circuitbasics.com/basics-uart-communication/)

UART to the sampling of bits by the receiving UART. The transmitting UART receives data in parallel from the data bus, then adds a start bit (i.e., a beginning of packet), a parity bit (i.e., a check error), and a stop bit (i.e., the end of a packet) to the data packet being transferred with the possibility of interruption during transmission. Therefore, when it detects a start bit, the receiving UART starts to read incoming bits at a specific frequency (i.e., baud rate or speed of data transfer in bits per seconds [bps]). The receiving UART discards the start bit, the parity bit, and the stop bit from the data frame. The receiving UART converts the serial data back into parallel and transfers it to the data bus on the receiving end. Both UARTs should operate at about the same baud rate, which can differ by about 10% before the timing of bits gets too far off. In addition, both UARTs should be configured to transmit and receive the same data packet structure.

*3.4.1.2 Serial Peripheral Interface (SPI) Protocol*  The Serial Peripheral Interface (SPI) is a synchronous communication protocol used by many different devices (e.g., SD card modules, radio frequency identification [RFID] card reader modules, etc.) to communicate with microcontrollers. In SPI, data is transferred in a continuous stream without interruption. The devices communicating via SPI protocol are in a master-slave relationship; where the master is the controlling device (i.e., a microcontroller), and the slave (i.e., a sensor, a display, or a memory chip) takes instruction from the master. This relationship could be between one master and one slave, as shown in Figure 3.2, as well as between one master and multiple slaves (i.e., the master has multiple slave select pins where the slaves can be wired in parallel [Figure 3.3], or the master has one slave select pin where the slaves can be daisy-chained [Figure 3.4]).

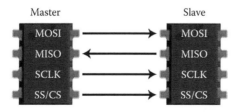

**Figure 3.2**    SPI single master and single slave connection diagram. (Source: Circuit Basics. 2016. Basics of UART Communication. http://www.circuitbasics.com/basics-uart-communication/)

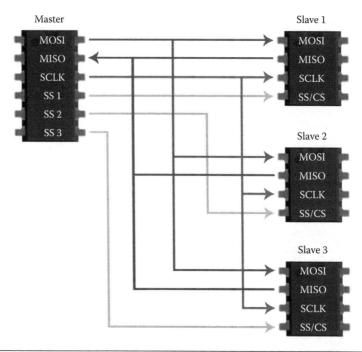

**Figure 3.3** Multiple slaves wired in parallel to single master in SPI. (Source: Circuit Basics. 2016. Basics of UART Communication. http://www.circuitbasics.com/basics-uart-communication/)

In SPI, we have four wires:

- Master Output/Slave Input (MOSI): This is a line for the master to send data to the slave.
- Master Input/Slave Output (MISO): This is a line for the slave to send data to the master.
- Clock (SCLK): This is a line for the clock signal.
- Slave Select/Chip Select (SS/CS): This is a line for the master to select to which slave to send data.

In SPI, the communication is synchronous and serial with a speed up to 10 Mbps. The clock signal synchronizes the output of data bits from the master to the sampling of bits by the slave. Therefore, one bit of data is transferred in each clock cycle, and the speed of data transfer is determined by the frequency of the clock signal.

Moreover, the SPI communication is always initiated by the master, since this configures and generates the clock signal. The master

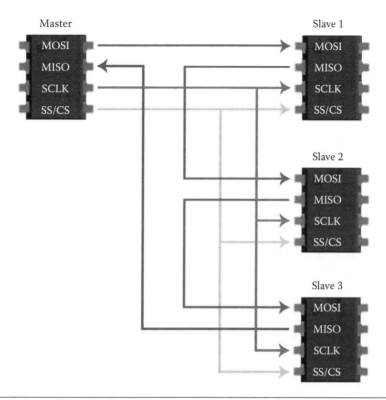

**Figure 3.4**    Multiple slaves daisy-chained to single master in SPI. (Source: Circuit Basics. 2016. Basics of UART Communication. http://www.circuitbasics.com/basics-uart-communication/)

outputs the clock signal, then in order to activate the slave, switches the SS/CS pin to a low-voltage state. The master sends the data one bit at a time to the slave along the MOSI line. The slave reads the bits as they are received. If a response is needed, the slave returns data one bit at a time to the master along the MISO line, then the master reads the bits as they are received.

*3.4.1.3 Inter-Integrated Circuit (I2C) Protocol*    An Inter-Integrated Circuit (I2C) is a synchronous and serial communication protocol (Circuit Basics 2016). In I2C, multiple slaves can be connected to single masters, and multiple masters can control single or multiple slaves. This will be useful when having multiple microcontrollers logging data to a single memory card or displaying text to a single LCD.

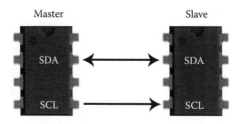

**Figure 3.5**   I2C Single master and single slave connection diagram. (Source: Circuit Basics. 2016. Basics of UART Communication. http://www.circuitbasics.com/basics-uart-communication/)

I2C uses two wires to transmit data between master and slave devices, as shown in Figure 3.5 (Circuit Basics 2016):

- Serial Data (SDA): This is a line for the master and slave to send and receive data.
- Serial Clock (SCL): This is a line that carries the clock signal controlled by the master, which synchronizes the output of bits to the sampling of bits between the master and the slave.

The maximum speed is equal to 100 kbps in standard mode, 400 kbps in fast mode, 3.4 Mbps in high-speed mode, and 5 Mbps in ultra-fast mode.

With I2C, the data is transferred in messages that are broken into frames of data. Each message has:

- An address frame containing the binary address of the slave composed of 7 or 10 bits
- One or more data frames containing the data being transmitted
- Start condition, where the SDA line switches from a high-voltage level to a low-voltage level before the SCL line switches from high to low
- Stop condition, where the SDA line switches from a low-voltage level to a high-voltage level after the SCL line switches from low to high
- Read/write bit to inform the slave whether the master wants to write data to it (read/write bit is a low-voltage level) or receive data from it (read/write bit is a high-voltage level)
- Acknowledgment/negative-acknowledgment (ACK/NACK) bits between each data frame to verify that the frame has been received successfully

In I2C data transmission, the master sends the start condition to every connected slave by switching the SDA line from a high-voltage level to a low-voltage level before switching the SCL line from high to low. Then, the master sends each slave, along with a read/write bit, the address frame of the slave it wants to communicate with. Each slave compares the address sent from the master to its own address. In case the address matches, the slave returns an ACK bit by pulling the SDA line low for one bit, otherwise, the slave leaves the SDA line high. After that, the master sends or receives the data frame. After each transferred data frame, the receiving device returns another ACK bit to the sender to acknowledge successful receipt of the frame. To stop the data transmission, the master sends a stop condition to the slave by switching SCL high before switching SDA high.

### 3.4.2 Comparison of the Different Communication Protocols of Microcontrollers

Table 3.2 represents a comparative study of UART, SPI, and I2C communication protocols of microcontrollers, as well as the strengths and the weaknesses of each (Circuit Basics 2016).

### 3.4.3 Internet Communication Protocols

New communication modules are oriented to work with Internet servers implementing internally several application layer protocols, as shown in Figure 3.6, which makes it easier to send the information to the cloud (Comer 2000). Among these application layer protocols, we can list the Hyper Text Transfer Protocol (HTTP), Hyper Text Transfer Protocol Secure (HTTPS), File Transfer Protocol (FTP), File Transfer Protocol Secure (FTPS), Simple Mail Transfer Protocol (SMTP), and Post Office Protocol 3 (POP3) (Alani 2014; Comer 2000; Edwards and Bramante 2009). These applications have a reliable transport by using the Transmission Control Protocol (TCP). The servers use well-known port numbers for the standardized services (e.g., port 80 for HTTP in the World Wide Web, port 443 for HTTPS, port 20 or 21 for FTP in file transfer, port 25 for SMTP in e-mail, port 110 for POP3 in remote e-mail access, etc.).

**Table 3.2** Comparison of Communication Protocols of Microcontrollers

| PROTOCOLS | STRENGTHS | WEAKNESSES |
|---|---|---|
| UART | Only uses two wires (i.e., Tx and Rx)<br>No Clock signal is required<br>Has a parity bit to check error<br>Structure of data packet can be changed when both sides are set up for it<br>Widely used method | Limited size of data to 9 bits<br>Does not support multiple slave or multiple master systems<br>Baud rates of each UART must be within 10% of each other |
| SPI | No start and stop bits (i.e., data is streamed continuously without interruption)<br>No complicated slave addressing system<br>High data transfer rate<br>Separate MISO and MOSI lines (i.e., data can be sent and received at the same time) | Uses four wires (i.e., MOSI, MISO, SCLK, and SS/CS)<br>No acknowledgment of receiving the data successfully<br>No parity bit to check the error<br>Only one single master |
| I2C | Only uses two wires (i.e., SDA and SCL)<br>Supports multiple masters and multiple slaves<br>ACK/NACK bit confirms the successful transfer of each frame<br>Less complicated hardware than UART<br>Widely used protocol | Slow data transfer rate<br>Limited size of data frame to 8 bits<br>More complicated hardware than SPI<br>Complicated slave addressing system |

TCP/IP Reference model

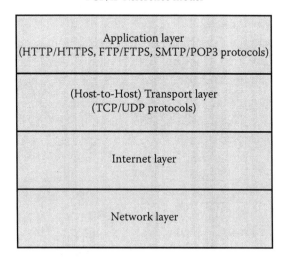

**Figure 3.6** TCP/IP reference model.

*3.4.3.1 Transmission Control Protocol (TCP)* A Transmission Control Protocol (TCP) is a full-duplex communication protocol at the level of the transport layer in the TCP/IP (i.e., Internet model) reference model, as shown in Figure 3.6. A TCP provides a considerable number of services to the Internet Protocol (IP) layer and to the application layer (Alani 2014; Comer 2000; Edwards and Bramante 2009). Also, it provides a connection-oriented protocol, which requires handshaking to set up end-to-end communications and enables an application to be sure that a packet sent out over the network was received in its entirety. TCP acts as a message-validation protocol providing reliable communications host-to-host. Due to network congestion, traffic load balancing, or other unpredictable network behavior, IP packets can be lost, duplicated, or delivered out of order. TCP detects these problems, requests retransmission of lost data, and rearranges out-of-order data. Once the TCP receiver has reassembled the sequence of octets originally transmitted, it passes them to the application program. Moreover, TCP helps minimize network congestion to reduce the occurrence of other problems. Therefore, TCP manages the flow of data streams from the application layer, as well as the incoming packets from the IP layer, by ensuring that priorities and security are respected. TCP uses port numbers to identify sending and receiving application end-points on a host, or Internet sockets. Each side of a TCP connection has an associated 16-bit unsigned port number (0-65535) reserved by the sending or receiving application. TCP is a protocol used by major Internet applications such as the World Wide Web, e-mail, remote administration, and file transfer.

*3.4.3.2 User Datagram Protocol (UDP)* The User Datagram Protocol (UDP) is a communication protocol at the level of the transport layer in the TCP/IP reference model, as shown in Figure 3.6. UDP provides a connectionless datagram protocol and offers no flow control (Alani 2014; Comer 2000; Edwards and Bramante 2009). As well, UDP is unreliable (i.e., no handshaking) using a weak checksum algorithm to detect error. Therefore, UDP messages can be lost, duplicated, or arrive out of order. Thus, the packets can arrive faster than the recipient can process them. On the other hand, the application has to provide its own error recovery, flow control, and congestion control functionality. UDP is used for multicasting and broadcasting, since retransmissions are not possible for a large amount of hosts. UDP typically gives higher throughput and shorter latency. It is therefore often used for real-time multimedia communication where packet loss occasionally can be accepted. It is also used for applications which do not require reliable data stream service and on-time arrival is more important than reliability (e.g., audio, video, Voice over IP, IP-TV, IP-telephony, online computer games, etc.), or for simple query/response applications like Domain Name System (DNS) lookups where the overhead of setting up a reliable connection is disproportionately large. Each side of a UDP connection has an associated 16-bit integer port number (0-65535) reserved by the sending or receiving application. Port 0 is reserved, but is a permissible source port value if the sending process does not expect messages in response. UDP provides application multiplexing (via port numbers) and integrity verification (via checksum) of the header and payload.

*3.4.3.3 Hypertext Transfer Protocol (HTTP) and Hypertext Transfer Protocol Secure (HTTPS)*   The Hypertext Transfer Protocol (HTTP) is a bidirectional communication protocol at the level of the application layer in the TCP/IP reference model, as shown in Figure 3.6, for distributed and collaborative hypermedia information systems (Alani 2014; Comer 2000; Edwards and Bramante 2009). HTTP is used for communication between a Web browser and a Web server or among immediate machines and Web servers. This protocol aims to exchange or transfer hypertext. Furthermore, HTTP functions as a request-response protocol in the client-server computing model. The client (e.g., Web browser) submits an HTTP request message to the server (e.g., an application running on a computer hosting a Web site). The server returns a response message to the client, which contains completion status information about the request and may also contain requested content in its message body. HTTP defines methods to indicate the desired action to be performed on the identified resource (e.g., GET request, to retrieve data; POST request, to post data to a database; etc.). Therefore, HTTP is stateless, where each request is self-contained, and the server does not keep a history of previous requests or previous sessions (i.e., a sequence of network request-response transactions). Furthermore, HTTP allows browsers and servers to negotiate details such as the character set to be used during transfers; where a sender can specify the capabilities it offers and a receiver can specify the capabilities it accepts. In addition, HTTP supports for caching (i.e., a browser caches a copy of each Web page it retrieves) to improve response time. In other words, if a user requests a page again, HTTP allows the browser to interrogate the server to determine whether the contents of the page have changed since the copy was cached. Moreover, HTTP supports intermediaries, which allows a machine along the path between a browser and a server to act as a proxy server that caches Web pages and answers a browser's request from its cache. HTTP assumes a reliable connection-oriented transport protocol (i.e., a TCP protocol) by establishing a connection to a particular port (i.e., port 80) on a server, but does not provide reliability or retransmission itself. However, HTTP can use unreliable protocols such as the User Datagram Protocol (UDP).

The Hypertext Transfer Protocol Secure (HTTPS) is a communication protocol for secure communication over a computer network, which is widely used on the Internet (Alani 2014; Edwards and Bramante 2009). HTTPS consists of communications over HTTP within a connection encrypted by Transport Layer Security (TLS) or Secure Sockets Layer (SSL) protocols. HTTPS provides an authentication of the visited Website and protection of the privacy and integrity of the exchanged data. Moreover, HTTPS provides bidirectional encryption of communications between a client and server, which protects against eavesdropping and tampering with or forging the contents of the communication by any third party. Therefore, HTTPS connections are used for payment transactions on the World Wide Web, e-mail, sensitive transactions in corporate information systems, protecting page authenticity on all types of Websites, securing accounts and keeping user communications, identity and web browsing private. HTTPS uses port 443.

*3.4.3.4 File Transfer Protocol (FTP) and File Transfer Protocol Secure (FTPS)*   The File Transfer Protocol (FTP) is a communication protocol at the level of the application layer in the TCP/IP reference model, as shown in Figure 3.6, which allows file transfer amongst computer users connected over a network (Alani 2014; Comer 2000; Edwards and Bramante 2009). It is a reliable, simple, and efficient protocol. FTP is built on a client-server architecture, uses TCP as a transport protocol to provide reliable host-to-host connections, and implements two types of connections in managing data transfers. FTP uses a separate control connection (i.e., referred to ephemeral port 21 to accept new connections) and data connection (i.e., referred to ephemeral port 20 for data transfer) between the client and the server. Therefore, FTP only opens a data connection when a client issues a command requiring a data transfer (e.g., request to retrieve a file, or to view a list of the files available). As well, it is possible for an entire FTP session to open and close without a data connection ever having been opened. The data connection is unidirectional, unlike the control connection in which commands and replies can flow both from the client to the server and from the server to the client. FTP can transfer data only from the client to the server, or from the server to the client, but not both. Also, unlike the control connection, the data connection can be initiated from either the client or the server, where data connections initiated by the server are active, and those initiated by the client are passive. Furthermore, FTP provides an interactive access, where most implementations provide an interactive interface that allows humans to easily interact with remote servers (e.g., a user can ask for a listing of all files in a directory on a remote machine, and a client usually responds to the input "help" by showing the user information about possible commands that can be invoked). Moreover, FTP allows the client to specify the type and format of stored data (e.g., a file contains text or binary integers, text files use the ASCII or EBCDIC character sets, etc.). In addition, FTP offers authentication control, which requires clients to authorize themselves by sending a login name and password to the server before requesting file transfers. The server refuses access to clients that cannot supply a valid login and password. FTP is considered an out-of-band protocol, as opposed to an in-band protocol (e.g., HTTP).

FTP was not designed to be a secure protocol and has many security weaknesses (e.g., brute force attacks, bounce attacks, packet capture [sniffing], port stealing, etc.) (Forouzan 2000). Username protection FTP is not able to encrypt its traffic. All transmissions are in clear text, and usernames, passwords, commands, and data can be easily read by anyone able to perform packet capture (sniffing) on the network. A common solution to this problem is to use the "secure" Transport Layer Security (TLS) or the Secure Sockets Layer

(SSL) protected version File Transfer Protocol Secure (FTPS) of the insecure protocol FTP. In other words, FTPS is an extension of the commonly used FTP protocol, which allows encrypting the session of FTP that adds support for the TLS and the SSL cryptographic protocols. FTPS implicit TLS/SSL services run on port 990.

*3.4.3.5 Simple Mail Transport Protocol (SMTP)*   The Simple Mail Transfer Protocol (SMTP) is a one-to-many communication protocol at the level of the application layer of the TCP/IP reference model, as shown in Figure 3.6 (Alani 2014; Comer 2000; Edwards and Bramante 2009). It is considered an Internet standard for electronic mail (e-mail) transmission across Internet Protocol (IP) networks. Therefore, electronic mail servers and other mail transfer agents use SMTP to send and receive mail messages. However, user-level client mail applications use SMTP only for sending messages to a mail server for relaying and use the Post Office Protocol (POP) (cf. Section 3.4.3.6) for receiving messages or other protocols to access their mail box accounts on a mail server. E-mail is submitted by a mail client (i.e., mail user agent [MUA]) to a mail server (i.e., mail submission agent [MSA]) using SMTP on TCP port 587, where most mailbox providers still allow submission on traditional port 25. Exchange Server 2013 has SMTP service listening on ports 25, 587, 2525, 465, and 475 depending upon server role and function. Once communication has been established, the sender can transmit one or more mail messages, terminate the connection, or request the server to exchange the roles of sender and receiver so messages can flow in the opposite direction. The receiver acknowledges each message. In addition, it can abort the entire connection or abort the current message transfer. SMTP does not specify the details of how a client handles such errors, the client must decide. SMTP allows the server to inform the client about the new address, which will use it in the future. When informing the client about a new address, the server may choose to forward the mail that triggered the message, or it may request that the client takes the responsibility for forwarding.

*3.4.3.6 Post Office Protocol (POP3)*   Post Office Protocol (POP3) is the most popular communication protocol at the level of the application layer of the TCP/IP reference model, as shown in Figure 3.6, used to transfer e-mail messages from a permanent mailbox to a local computer (Alani 2014; Comer 2000; Edwards and Bramante 2009). The user invokes a POP3 client, which creates a TCP connection to a POP3 server on the mailbox computer. The user first sends a login and a password to authenticate the session. Once authentication has been established, the user client sends commands to retrieve a copy of one or more messages and to delete the message from the permanent mailbox or to keep a copy of the message for a predefined number of days. Furthermore, the computer with the permanent mailbox run two servers: an SMTP server to accept mail sent to a user and add each incoming message to the user's permanent mailbox, and a POP3 server to allow a user to extract messages from the mailbox and delete them. To ensure correct operation, it is required that the two servers coordinate when using the mailbox; so that if a message arrives via SMTP while a user is extracting messages via POP3, the mailbox is left in a valid state. POP3 uses the port 110 and the port 995 when connecting securely.

### 3.4.4 Comparison of Different Internet Communication Protocols

Table 3.3 represents a comparative study between TCP and UDP transport layer communication protocols (Alani 2014; Comer 2000; Edwards and Bramante 2009), and Table 3.4 represents a comparative study and the advantages and disadvantages among HTTP, HTTPS, FTP, FTPS, SMTP, and POP3 application layer Internet communication protocols (Alani 2014; Comer 2000; Edwards and Bramante 2009; Forouzan 2000).

**Table 3.3**   Comparison of TCP and UDP Transport Layer Communication Protocols

| SPECIFICATIONS | TCP | UDP |
|---|---|---|
| Message Data Name | Packet | Datagram |
| Consistency | Reliable | Unreliable |
| Connection Type | Connection-oriented | Connectionless |
| On Error of Receiving | Segment retransmission and flow control through windowing | No segment retransmission and no flow control through windowing |
| Traffic Network | Congestion control | No congesting control |
| Ordering | Segment sequencing | No sequencing |
| Request/Response | Acknowledgement | No acknowledgment |

**Table 3.4** Comparison of Application Layer Communication Protocols

| SPECIFICATIONS | HTTP | HTTPS | FTP | FTPS | SMTP | POP3 |
|---|---|---|---|---|---|---|
| Applications | Navigation, downloading, and uploading content to a Web server | Navigation, downloading, and uploading content to a Web server securely | Transfer of files amongst computer users connected over network | Transfer of files amongst computer users connected over network securely | Transmission of electronic mail (e-mail) across Internet Protocol (IP) networks | Transfer e-mail messages from a permanent mailbox to a local computer |
| Port | 80 | 443 | 20 and 21 | 990 | 25 | 110 |
| Transport Protocol Used | TCP | TLS/SSL | TCP | TLS/SSL | TCP | TCP |
| Communication | Bidirectional | Bidirectional | Bidirectional | Bidirectional | Unidirectional one-to-many | Unidirectional one-to-one |
| Advantages | Request response  Capability negotiation  Support for caching  Support for intermediaries | Encrypted communications data  Request response  Capability negotiation  Support for caching  Support for intermediaries  Secure mode | Interactive access  Format representation facilities (e.g., text files use ASCII or EBCDIC  Required authentication | Encrypted communications data  Interactive access  Format representation facilities (e.g., text files use ASCII or EBCDIC  Required authentication  Secure mode | Ability to relay messages from one server to another when the sender and recipient have different e-mail service providers  E-mails can be sent to one person or several people | Required authentication to prevent malicious individuals from gaining unauthorized access to users' messages  E-mails can be sent to one person or several people |
| Disadvantages | Nonsecure mode  Stateless  Unreliable | Stateless  Unreliable | Nonsecure mode  Trivial | Trivial  Viruses are easily spread via e-mail attachments | Nonsecure mode | Nonsecure mode  Viruses are easily spread via e-mail attachments |

3.5 Conclusion

In this chapter we have detailed the various communication modules in the GPS system with the functionalities and the accuracy of each. In addition, we described the various communication protocols of microcontrollers. These protocols each have strengths and weaknesses as shown in Table 3.2. This comparative study justifies our choice for the communication protocols of microcontrollers. Internet communications protocols with the advantages and the disadvantages of each are integrated in this chapter to show the possible communication services and purpose of application between the GPS unit and the end user.

We consider that communication protocols of microcontrollers are important to communicate with the different peripherals, especially with the Integrated Development Environment (IDE) computer to transmit the developed programs to the microcontroller in specific programming languages. The next chapter will take this approach.

# 4

# PROGRAMMING
# MICROCONTROLLER

## 4.1 Introduction

The originality of the microcontroller is that it is a fully program-mable circuit capable of calculating and processing analog and logic signals present on its inputs and sending the results of the process-ing on the output circuits (Cazaubon 1997). This flexibility leads to making the microcontroller able to manage the most diverse applica-tions. Therefore, it is necessary to choose the resources and the means for implementation in order to develop an application (Cazaubon 1997; Peatman 1988). At the first stage, as the first source, it is cru-cial to choose the microcontroller according to the application to be processed (e.g., capacity of ROM and RAM, number of inputs and outputs, serial links, converters, etc.); then, as the second source, the possibilities of storage, as they are important elements in the choice of a component. In a second step, it is essential to verify that the selected microcontroller is supported by the available development system. In a final step, it is necessary to go to the writing phase of the program with a text editor according to the syntax of the assembler. The writing of the program involves many subprograms or general modules. This type of writing saves valuable time as soon as we have a well-structured program library.

Developing application software for a microcontroller requires development tools. While the chip itself is small and inexpensive, the tools needed to make it useful are expensive. Since the microcon-troller only understands the binary language (i.e., machine language), any program has to be loaded into memory in binary code (i.e., object program) in order to be executable by the microcontroller (Cazaubon 1997; Peatman 1988). Therefore, the assembler is software that allows translating the source file to an object file and is executable by the microcontroller. The assembler detects syntax errors and can create object files in several formats. Moreover, it is required to have a simulator or an emulator used for debugging a program. The simulator allows, under the control of the simulation software, to check that the program is proceeding correctly and taking into account the inputs and the outputs. On the other hand, the emulator allows, as the simulator, to check the correct execution of the program, but it verifies the good integration of the software for the application and its performance. In addition, the emulator fully simulates the microcontroller of the final application. In other words, the emulator interacts with the microcontroller. Moreover, the essential part of the emulator is the Electrically Erasable Programmable Read-Only Memory (EEPROM) or flash memory, which is switched by the RAM. Therefore, when the new data is written in the RAM, the emulator replaces the previous data. Recent microcontrollers are often integrated with on-chip debugging circuitry that, when accessed by an in-circuit emulator (ICE) via JTAG, allow debugging of the firmware with a debugger (Mazzei et al. 2015). A real-time ICE may allow viewing and/or manipulating of internal states while running. A tracing ICE can record executed programs and microcontroller states before/after a trigger point. Among the features of the emulator, we can list (Agarwal 2017; Peatman 1988):

- Understanding of Hex file format.
- Checking whether the data has burned to the RAM and the complete RAM may be read from the parallel port.
- Allowing two kinds of real time debugging: single step debugging where the user may implement the code and in each step only one instruction is executed, and breakpoint debugging in case the length of the code is long and then the user has one option to choose breakpoints at the location where he wants to implement them in one step.
- Updating, since after every instruction execution, there will be an update of the internal memory window of the microcontroller using a serial port.
- Using Command buttons to read the contents of the timers, register banks, ports, internal memory, and Special Function Registers (SFRs), which are provided in the menu bar.
- Knowing the execution of the breakpoints based on different color combinations.

## 4.2 Development System of Microcontroller Applications

The typical development system of microcontroller applications is composed of a computer to write a given application program (e.g., program to send text message containing the geolocation of a given object). This computer is linked to the microcontroller with an interface FTDI chipset (cf. Chapter 2, Section 2.5) using an UART communication protocol (cf. Chapter 3, Section 3.4.3.1) or programmer (cf. Chapter 2, Section 2.6) using a SPI communication protocol (cf. Chapter 3, Section 3.4.3.2). The application program is written in a high-level language (e.g., Embedded C, Python, JavaScript, etc.), compiled, converted into a machine code, and then sent for storage in the flash memory of the microcontroller through the FTDI chip using an UART serial communication or through the programmer using a SPI communication (Bilal 2017).

Each microcontroller has its own specific machine language (Gibilisco and Doig 2017). It is designed as a set of instructions. It reads and handles a certain number of bits at a time (i.e., an instruction) where its result forms an operation (e.g., arithmetic calculation, storing data in memory, etc.). Then it reads the next instruction, and so on.

Each microcontroller has its own manufactured Integrated Development Environment (IDE) and its own compiler. Microcontrollers were originally programmed only in assembly language, but various high-level programming languages (e.g., embedded C, Python, JavaScript, etc.) are now in common use to target microcontrollers and embedded systems (Mazzei et al. 2015).

Typically, microcontroller programs have to fit in the available on-chip memory, since it would be costly to provide a system with external, expandable memory (Bilal 2017). Compilers and assemblers are used to convert both high level and assembly language codes into a machine code for storage in the microcontroller's memory. Depending on the device, the program memory may be permanent, read-only memory (ROM) that can only be programmed at the factory, or it may be field-alterable flash. It could also be Electrically Erasable Programmable Read-Only Memory (EEPROM), which is easier to use and cheaper to manufacture. For example, embedded C that is used for microcontroller-based applications does not take more computer resources (e.g., memory, operating system, etc.).

The microcontrollers provide real-time response to events in the embedded system they are controlling (NewbieHack 2014). When certain events occur, an interrupt system can signal the processor to suspend processing the current instruction sequence and to begin an interrupt service routine (ISR) or "interrupt handler," which will perform any processing required based on the source of the interrupt, before returning to the original instruction sequence. Possible interrupt sources are device dependent, and often include events such as an internal timer overflow, completing an analog to digital conversion, a logic level change on an input (e.g., from a button being pressed and data received on a communication link), a counter's number, and changing a pin state from low to high or high to low. Moreover, since the power consumption is as important as in battery devices, the interrupts wake a microcontroller from a low-power sleep state, where the processor is halted until required to do something by a peripheral event (Atmel 2012). However, interrupts cannot be enabled during a programming session, since an interrupt will change the timing causing the memory cell to be altered (Svendsli 2003).

In analyzing problems or debugging programs, a tool called a dump is commonly used as a printout to show the program in its machine code form (Gibilisco and Doig 2017). However, each four bits are represented by a single hexadecimal numeral, since it is difficult to render the program entirely as zeros and ones.

### 4.3 Programming Languages

The programming languages of the microcontroller are classified into two categories: the low-level programming language and the high-level programming language. Low-level programming languages are specific for each microcontroller architecture (Cockerell 1987). However, most high-level programming languages are generally portable across multiple architectures, but require interpreting or compiling (Cockerell 1987). We are interested in the assembly language as a low-level programming language, and the C language and the Lua language as high-level programming languages.

#### 4.3.1 Assembly Language

An assembly language is a low-level programming language for a computer, or other programmable device (e.g., microcontroller), in which there is a very strong one-to-one correspondence between the language and the architecture's machine code instructions (Blum 2006; Cockerell 1987). Each assembly language is specific to a particular microcontroller architecture. Assembly language is also called symbolic machine code. It is converted into executable machine code by an assembler as a utility program. A program written in assembly language consists of a series of machine code instructions (i.e., mnemonic), meta-statements (i.e., directives, pseudo-instructions, and pseudo-ops), comments, and data. Assembly language instructions usually consist of an opcode mnemonic followed by a list of data, arguments, or parameters. These are translated by an assembler into machine language instructions that can be loaded into memory and executed. Multiple sets of mnemonics or assembly-language syntax may exist for a single instruction set, typically represented in different assembler programs. Therefore, they are supplied by the manufacturer and used in its documentation.

### 4.3.2 C Programming Language

C is a general-purpose, imperative computer programming language and high-level language originally developed by Dennis M. Ritchie to develop the UNIX operating system at Bell Labs in 1972 (Fresh2Refresh 2017; Tutorialpoints 2017). It is easy to learn, and it produces efficient programs. It was designed to be compiled using a relatively straightforward compiler, to provide low-level access to memory, to provide language constructs that map efficiently to machine instructions, and to require minimal run-time support. Therefore, C was useful for many applications that had formerly been coded in assembly language (e.g., system programming), including operating systems, as well as various application software for computers ranging from supercomputers to embedded systems (i.e., microcontrollers). It can handle low-level activities and can be compiled on a variety of computer platforms. Despite its low-level capabilities, the C language was designed to encourage cross-platform programming. C can be used for structured programming and allows lexical variable scope and recursion. C has now become a widely used professional language with C compilers from various vendors available for the majority of existing computer architectures and operating systems.

### 4.3.3 Lua Programming Language

Lua is a lightweight multi-paradigm scripting and high-level programming language designed in 1993 primarily for embedded systems and clients (Ierusalimschy et al. 1996, 2007). Lua is cross-platform, since it is written in ANSI C, and has a relatively simple C API. Lua is a proven, robust, embeddable extension language (i.e., strong integration with code written in other languages) (Lua 2011). It is portable, extensible, and has ease-of use in development with high speed in the realm of interpreted scripting languages and in execution (Cormack 2013; Ierusalimschy et al. 1996, 2007). It provides the basic facilities of most procedural programming languages and powerful data description by using a simple mechanism of tables. However, more complicated features (e.g., domain-specific features) are not included, rather, it includes mechanisms for extending the language, allowing programmers to implement such features. Lua is a dynamically typed language intended for use as an extension or scripting language and is compact enough to fit on a variety of host platforms. Therefore, Lua is provided as a library of C functions to be linked to host applications. It supports only a small number of atomic data structures (e.g., Boolean values, numbers, and strings). Typical data structures (e.g., arrays, sets, lists, and records) are represented using Lua's single native data structure. Moreover, Lua allows programmers to implement namespaces, classes, and other related features using its single table implementation. Therefore, Lua strives to provide simple, flexible meta-features that can be extended as required. In addition, Lua is a language framework, which copes with different domains, thus creating customized programming languages sharing a single syntactical framework. On the other hand, it is very easy to write an interactive, standalone interpreter for Lua. Although Lua does not have a built-in concept of classes, object-oriented programming can be achieved by using two language features: first-class functions and tables. Lua programs are not interpreted directly from the textual Lua file, but are compiled into bytecode, which is then run on the

Lua virtual machine. The compilation process is typically invisible to the user and is performed during run-time, but it can be done offline in order to increase loading performance or reduce the memory footprint of the host environment by leaving out the compiler. Lua bytecode can also be produced and executed from within Lua, using the dump function from the string library and the load/loadstring/loadfile functions. In case of programming a microcontroller with Lua, there is no need for Lua IDE to write a Lua script code, thus it is written in any text editor (e.g., Notepad ++, Ultraedit, etc.), then it saves in Lua format (i.e., extension.lua). After that, it will be downloaded into the EFS of the module to be compiled (i.e., extension.out) and executed (Libelium 2017). The eLua is used in programming microcontrollers and embedded systems where it is not a cutdown version (Cormack 2013).

## 4.4 Recapitulative Table of Comparison of Different Microcontroller Programming Languages

Table 4.1 represents a comparative study among different microcontroller programming languages by showing the differences, as well as the advantages and the disadvantages of each language (Blum 2006; Cockerell 1987; Cormack 2013; Fresh2Refresh 2017; Ierusalimschy et al. 1996, 2007; Lua 2011; Tutorialpoints 2017).

## 4.5 ATtention (AT) Command

The Hayes command set, known as the ATtention (AT) Standard command set, is a specific command language originally developed by Dennis Hayes for the Hayes Smart modem 300 baud modem in 1981 (Dalakov 1999; Nursat 2010). The Hayes Standard AT command set provides a solution to allow any computer using a serial port to control the modem functions with software (e.g., HyperTerminal) (Boyce 2002). Moreover, the two-character abbreviation, AT, is used to start a command line to be sent from Terminal Equipment/Data Terminal Equipment (TE/DTE) (e.g., computer) to Terminal Adaptor/Data Communication Equipment or Data Circuit terminating Equipment (TA/DCE) (e.g., data card, modem, etc.). The command set consists of a series of short text strings, which can be combined to produce

**Table 4.1** Comparison of Different Microcontroller Programming Languages

| SPECIFICATIONS | ASSEMBLY | C | LUA |
|---|---|---|---|
| Technologies | Low-level programming language | High-level programming language | High-level programming language |
| Correspondence | One-to-one correspondence between the language and the architecture's machine code instructions. High-level machine language | One-to-many correspondence between the language and the architecture's machine code instructions. High-level machine language | One-to-many correspondence between the language and the architecture's machine code instructions. High-level machine language |
| Instruction Statements | Very simple | Complicated | Complicated |
| Tool of Conversion to Machine Code | Assembler | Compiler | Interpreter |
| Advantages | Multiple sets of assembly-language syntax may exist for a single instruction set, typically represented in different assembler programs<br><br>No need for IDE | Easy to learn<br>Widely used professional language<br>Fast<br>Portable<br>Structured language<br>Produce efficient programs<br>Can handle low-level activities<br>Can be compiled on a variety of computer platforms<br>Cross-platform programming | Cross-platform programming<br>Robust, embeddable extension language<br>Portable<br>Extensible<br>Scripting language<br>Ease-of-use in development with high speed<br>Facilities of most procedural programming languages and powerful data description<br>No need for IDE<br>Use any text editor (e.g., Notepad++, etc.) to write the script code |
| Disadvantages | Not easy in development | IDE specific for each microcontroller provided by the manufacturer | Not include the more complicated features (e.g., domain-specific features) |

commands for operations such as dialing, hanging up, and changing the parameters of the connection. The vast majority of dial-up modems use the Hayes command set in numerous variations. Moreover, the command set covered only those operations supported by the earliest 300 bit/s modems. Furthermore, vendors provide a variety of one-off standards when new commands were required to control additional functionality in higher speed modems. Nowadays, these vendors re-standardize on the Hayes extensions due to the introduction of the high-speed modems.

## 4.6 NMEA Sentence and GPRMC Sentence

The National Marine Electronics Association (NMEA) has developed a specification that defines the interface between various pieces of marine electronic equipment (DePriest 2013; Leadtek 2012). GPS receiver communication is defined within this specification. The GPS data includes the complete PVT (position, velocity, and time) solution computed by the GPS receiver. The GPS data is represented in NMEA sentence and recommended minimum data for GPS (RMC) sentence format. Each of these sentences are represented as a line of data. Each GPS receiver has its standard sentence with the possibility to define custom sentences for use by a company. The standard sentences have a two letter prefix that defines the device using that sentence type (e.g., prefix GP for a GPS receiver). Also, the custom sentences begin with the letter P and are followed by three letters that identify the manufacturer controlling that sentence (e.g., PGRM for a Garmin sentence, PMGN for Magellan, etc.). Each sentence begins with 'S sign, ends with a carriage return/line feed sequence, can be no longer than 80 characters, and the data is separated by commas to determine the field boundaries. Most of the devices are designed to meet the NMEA requirements. We are interested in NMEA sentences and the GPS recommended minimum data for a GPS (GPRMC) sentence. The RMC—NMEA has its own version of essential GPS PVT data called RMC. A sample of RMC is represented as the following:

$GPRMC,123519,A,4807.038,N,01131.000,E,022.4,084.4,230394, 003.1,W*6A

Where,

| | |
|---|---|
| RMC | Recommended Minimum sentence C |
| 123519 | Fix taken at 12:35:19 UTC |
| A | Status A = active or V = Void. |
| 4807.038,N | Latitude 48 degrees 07.038′ N |
| 01131.000,E | Longitude 11 degrees 31.000′ E |
| 022.4 | Speed over the ground in knots |
| 084.4 | Track angle in degrees |
| 230394 | True Date - 23rd of March 1994 |
| 003.1,W | Magnetic Variation |
| *6A | The checksum data, always begins with* |

### 4.7 Conclusion

In this chapter we have detailed the development system of microcontroller applications. In addition, we developed the various programming languages of the microcontroller with the advantages and the disadvantages of each in order to justify our choice. The Attention (AT) command is integrated to allow the facilities of any computer using a serial port to control the modem functions with software, such as HyperTerminal. Similarly, the NMEA sentence and the GPRMC sentence are presented to show the format of the GPS information posted to the Web. These sentences are sent to the Web to be written in the database installed on the server or via SMS.

Our discussion will turn to conceiving an efficient model of the tracking system for the fleet management. This model will be detailed in the next chapter.

# 5

# OUR PROPOSED MODEL: INFELECPHY GPS UNIT

## 5.1 Introduction

The objective of the proposed model InfElecPhy GPS Unit (IEP-GPS) is to do the tracking for users of fleet management. This model is based on the dedicated 3G/GPRS module (cf. Chapter 3, Section 3.2.2) to provide efficient information to the end user. This model is founded on two architectures: the use of a microcontroller and not using a microcontroller to do the tracking for users. However, the model without a microcontroller requires less resources, and it is less costly and with a reduced size as compared to the model with the microcontroller, even if the microcontroller has evidently been efficient in resolving several problems in GPS devices. The IEP-GPS model provides reliability since it deals with several protocols (cf. Chapter 3, Section 3.4.3): 1) HTTP and HTTPS to navigate, download, and upload in real time the information to a Web server, 2) FTTP and FTTPS to handle in a non-real time the files to the Web application, and 3) in case of any alert, SMTP and POP3 to send and receive e-mail directly from the unit. Moreover, this model is similar to a mobile device, but without a screen for display. It is multifunctional, since it will be linked to a 3G/GPRS module, a camera, a speaker, a headphone, a keypad, and a screen. Furthermore, this model will provide accuracy in the tracking process, since it will work when we do not have a connection to the network, to store on a SD card socket the necessary tracking information, and it will be sent automatically once connected to the Internet. In addition, the IEP-GPS model is compatible with the open source application, OpenGTS (Geotelematic 2015), to do tracking for given users on various types of maps and to edit various reports. Moreover, this model will provide the facility to communicate with the application OpenGTS installed locally on the server of the client's company and not the provider due to the limitation of Internet usage. Therefore, our proposed model aims to improve the existing models for the storage of the information when disconnection occurs from the network. Our model also aims for adaptability with the open source applications (e.g., OpenGTS) to gather the diverse required information. In addition, our model aims to edit various reports with zero cost of development. It has flexibility to install the application OpenGTS on the local server of the company to retrieve the required information in a quick way due to the limitations of Internet usage.

### 5.2 Proposed Model InfElecPhy GPS Unit (IEP-GPS)

Our proposed model InfElecPhy GPS Unit (IEP-GPS) does the tracking for the users of fleet management. This model is based on a dedicated GPRS/3G module to provide accurate tracking information to the end user.

This model is founded on two architectures: the use of the microcontroller and not using the microcontroller. With the use of microcontroller architecture, it is more effective to use the AVR microcontroller with a small 8-bit bus width which reduces the overall power consumption and the cost. This is faster than the other types of microcontrollers (cf. Chapter 2, Section 2.4). Also, the Future Technology Devices International (FTDI) chipset is faster than the programmer to program the microcontroller (cf. Chapter 2, Section 2.5 & Section 2.6). Moreover, the microcontroller and the GPS module are placed on two separated chipsets, which require more size in the GPS unit than the one without use of the microcontroller architecture. In both architectures, at the level of the GPS module, the GPRS/3G module and the GPS receiver are placed on the same chipset separately from the SD card. At the level of the GPS module, we place the SD card in the middle between the GPRS/3G module and the GPS receiver for better design.

The Lua scripting language is more robust, so there is no need for specific Interface Development Environment (IDE) to write the script code, and it has ease-of-use in development with high speed than do the other programming languages (cf. Chapter 4, Section 4.4).

This model should provide accuracy in the tracking process, especially when disconnection occurs from the network. As well, this model allows obtaining a tracking of the users on various types of maps and editing various reports with zero cost of development. Moreover, this model provides a solution to the limitation of Internet usage when collecting tracking data for analysis and reporting.

### 5.2.1 *Architecture of the Model IEP-GPS with Microcontroller*

The architecture of the model IEP-GPS with the microcontroller is composed of the following:

- A computer PC (IDE) that contains the proper Integrated Development Environment (IDE) (i.e., C programming language) and the proper compiler for the implemented microcontroller. In addition, this computer is able to display the information received from the microcontroller.
- A FTDI chipset (cf. Chapter 2, Section 2.5) converter from serial to USB in order to transfer the developed programs from the PC (IDE) to the memory of the microcontroller and to transfer the debugging information from the microcontroller to the PC (IDE) for display on the interface of the IDE installed on the PC (IDE).
- A microcontroller of type AVR (cf. Chapter 2, Section 2.3.4) that has 8 bit as bus width; one clock per instruction cycle as speed; SRAM and FLASH as memory; RISC as instruction set architecture; and low-power consumption. This microcontroller contains a bootloader (cf. Chapter 2, Section 2.7) to be executed at first when the microcontroller is initialized. Also, this microcontroller includes a flash memory that contains all the programs of the microcontroller developed on the PC (IDE).
- A GPS module that contains three components: a GPRS/3G module (cf. Chapter 3, Section 3.2.2) to communicate data from the microcontroller to the concerned controller user along with SMS, or to the Web along with TCP, HTTP, FTP or other services (cf. Chapter 3, Section 3.4.3); a SD card to store up to 32 GB of information (e.g., GPS information when disconnection occurs from the network, etc.) (cf. Chapter 2, Section 2.9); and a GPS receiver to get the signal

from the satellite (cf. Chapter 3, Section 3.2.3). In our case, we get signals from four satellites in order to have an accurate position (cf. Chapter 1, Section 1.1).

- A GPS antenna is connected to the GPS receiver to boost the GPS signal (cf. Chapter 2, Section 2.10). The GPS antenna should be active with good visibility to the sky, high gain, have a low noise amplifier and filter (cf. Chapter 2, Section 2.10).
- A GSM antenna to establish the connection between the GPRS/3G module and the Internet PC along with HTTP, TCP, or FTP services; or between the GPRS/3G module and the controller user along with SMS service.
- An Internet PC that receives the data from the GPRS/3G module and to be posted on the database "DB Client/Server" that is installed on the Server PC. The connection is established through a fixed broadcast IP. This database contains information about the controller users, the users for tracking (i.e., a username, a password, and a department), information about the tracking of the fleet management of the tracking users (e.g., time, status, latitude, longitude, speed, date, etc.), and alert messages received by e-mail.
- The controller user is the user that controls the tracking for the users of the fleet management. The controller user is identified by his mobile number predefined in the SMS application program installed on the microcontroller in order to receive alert messages.

This model can handle a GPS camera (cf. Chapter 2, Section 2.11) connected to the GPS module in order to take a snapshot photo, for example, when the user is out of the geofence. This photo will be sent via e-mail to the controller user. In addition, this camera can support video calls.

In this model, the communication between the microcontroller and the peripheral components (i.e., FTDI, GPS module), as well as between the FTDI and the PC (IDE), is a USB-to-serial using the Universal Asynchronous Receiver/Transmitter (UART) protocol (cf. Chapter 3, Section 3.4.3.1).

Each developed program in C (e.g., a program to send a text message containing the geolocation of a given object, a program to take a location and send it by e-mail, a program to send an alert text SMS message to the controller user, etc.) on the PC (IDE) will be compiled using the proper compiler for the microcontroller. This is converted into a machine code, and then sent for storage in the flash memory of the microcontroller (or the bootloader in case of an initialization program for the microcontroller) through the FTDI chipset using UART serial communication (cf. Chapter 4, Section 4.2). Also, this developed program can be debugged, and the result of the debugging will be sent from the microcontroller to the IDE interface for display.

The GPS receiver sends the GPS information received from the satellite through the satellite signal and after conversion to NMEA sentence format (cf. Chapter 4, Section 4.6). The microcontroller sends an AT command (cf. Chapter 4, Section 4.5) to the GPRS/3G module to check the connectivity along with an UART protocol, and to start the data transmission over the Internet by using SMS, TCP, HTTP, FTP or other services. In case of disconnection from the network, the microcontroller sends the location information for storage in the SD card while not connected, then it will be transmitted to the concerned destination once the connection to the network is back. The transmission of this stored information is issued in idle time, since the time to send the GPS information is predefined (e.g., 2 minutes).

Each user for tracking is defined by a username, which is assigned to his proper IEP-GPS unit as an identifier. The information of this user (i.e., a username, a password, and a department) is saved in the database installed on the DB Client/Server.

Figure 5.1 shows the overall architecture of the proposed model IEP-GPS with a microcontroller.

*5.2.2 Architecture of the Model IEP-GPS without a Microcontroller*

The architecture of the model IEP-GPS without a microcontroller is composed of the following:

- A computer PC (Lua) that contains the text editor Notepad++, Lua scripting programming language, and its proper EFS to compile the Lua scripting code (cf. Chapter 4, Section 4.3.3) for the GPS module. As well, the information received from the GPS module (e.g., tracing information at each step of the

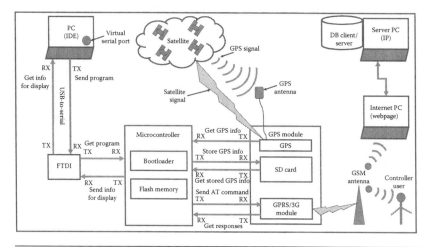

**Figure 5.1**   Architecture of the proposed model IEP-GPS with a microcontroller.

code, response of AT command sets, etc.) is displayed on the HyperTerminal software interface (Boyce 2002) implemented on this computer.

- A GPS module acting as microcontroller to communicate with the different peripheral components of this model. This GPS module contains three components: A GPRS/3G module (cf. Chapter 3, Section 3.2.2) to communicate data from the GPS module to the concerned controller user along with SMS, or to the Web along with TCP, HTTP, FTP or other services (cf. Chapter 3, Section 3.4.3); a SD card to store up to 32 GB of information (e.g., GPS information when disconnection occurs from the network, etc.) (cf. Chapter 2, Section 2.9); and a GPS receiver to get the signal from the satellite (cf. Chapter 3, Section 3.2.3). In our case, we get signals from four satellites in order to have an accurate position (cf. Chapter 1, Section 1.1).

- A GPS antenna is connected to the GPS receiver to boost the GPS signal (cf. Chapter 2, Section 2.10). The GPS antenna should be active with good visibility to the sky, high gain, and have a low noise amplifier and filter (cf. Chapter 2, Section 2.10).

- A GSM antenna to establish the connection between the GPRS/3G module and the Internet PC along with HTTP, TCP, or FTP services; or between the GPRS/3G module and the controller user along with SMS service.

- An Internet PC that receives the data from the GPRS/3G module and posts on the database "DB Client/Server" installed on the Server PC. The connection is established through a fixed broadcast IP. This database contains information about the controller users, the users for tracking (i.e., a username, a password, and a department), information about the tracking of the fleet management of the tracking users (e.g., time, status, latitude, longitude, speed, date, etc.), and alert messages received by e-mail.
- The controller user is the user that controls the tracking for the fleet management of the given users. The controller user is identified by his mobile number predefined in the SMS application program installed on the GPS module in order to receive alert messages.

This model can handle the GPS camera (cf. Chapter 2, Section 2.11) connected to the GPS module in order to take a snapshot photo, for example, when the user is out of the geofence. This photo will be sent via e-mail to the controller user. As well, this camera can support video calls.

In this model, the communication between the GPS module and the PC (Lua) is a USB-to-serial using the Universal Asynchronous Receiver/Transmitter (UART) protocol (cf. Chapter 3, Section 3.4.3.1).

Each developed scripting code in Lua (e.g., a program to send text message containing the geolocation of a given object, a program to take a location and send it by e-mail, and a program to send alert text SMS messages to the controller user, etc.) on the PC (Lua) will be compiled using the proper EFS for the Lua (cf. Chapter 4, Section 4.3.3), and then sent for storage in the GPS module using UART serial communication (cf. Chapter 4, Section 4.2).

The GPS receiver sends the GPS information received from the satellite through the satellite signal and after conversion to NMEA sentence format (cf. Chapter 4, Section 4.6). The GPS module sends an AT command (cf. Chapter 4, Section 4.5) to the GPRS/3G module to check the connectivity, and to start the data transmission over the Internet by using SMS, TCP, HTTP, FTP or other services. In case of disconnection from the network, the GPS module sends

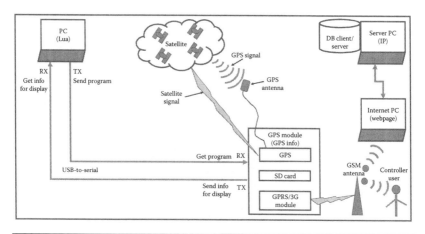

**Figure 5.2**    Architecture of the proposed model IEP-GPS without a microcontroller.

the location information for storage in the SD card while not connected, then it will be transmitted to the concerned destination once the connection to the network is back. The transmission of this stored information is issued in idle time, since the time to send the GPS information is predefined (e.g., 2 minutes).

Each user for tracking is defined by a username, which is assigned to his proper IEP-GPS unit as an identifier. The information of this user (i.e., a username, a password, and a department) is saved in the database installed on the DB Client/Server.

Figure 5.2 shows the overall architecture of the proposed model IEP-GPS without a microcontroller.

### 5.3 Representation of Interactions between Different Actors of the Model IEP-GPS

The interactions between the different actors of the conceived model IEP-GPS with a microcontroller and without a microcontroller are described, respectively, in Section 5.3.1 and Section 5.3.2.

#### 5.3.1 Interactions between Different Actors of the Model IEP-GPS with a Microcontroller

The interactions between the different actors of the conceived model IEP-GPS with a microcontroller are described in the sequence

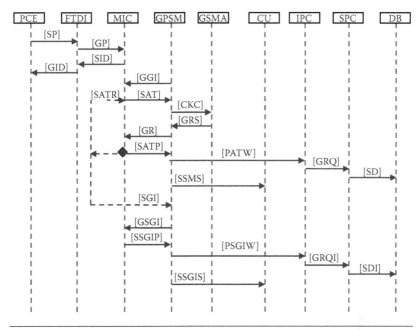

**Figure 5.3**    Interactions between the different actors of the model with a microcontroller.

diagram shown in Figure 5.3. The notations used in Figure 5.3 are as follows:

PCE        Computer PC (IDE)
FTDI       FTDI chipset
MIC        Microcontroller
GPSM       GPS module
GSMA       GSM antenna
CU         Controller user
IPC        Internet PC to access the Webpage
SPC        Server (PC)
DB         Database client/server
[SP]       [Send program to configure the microcontroller or for initiation]
[GP]       [Get program received from PC (IDE)]
[SID]      [Send information about program's debugging for display]
[GID]      [Get information about program's debugging for display]
[GGI]      [Get GPS information]
[SAT]      [Send AT command to check the network connectivity]
[CKC]      [Check the network connectivity]

[GRS]     [Get GPS response signal from GSM antenna about network connectivity]

[GR]      [Get response about network connectivity]

[SATP]    [If response is positive, send AT command to be parsed on the Web or via SMS]

[PATW]    [Parse AT command over the web]

[GRQ ]    [Get request from Internet PC about parsing of information received from the GPRS/3G module]

[SD]      [Store parsing data]

[SSMS]    [Send text SMS message to the controller user]

[SATR]    [Repeat sending AT command to check the network connectivity]

[SGI]     [If no network connection, send GPS information for storage in the SD card]

[GSGI]    [If network connection is back, get stored GPS information back from SD card]

[SSGIP]   [Send stored GPS information to be parsed on the Web or via SMS]

[PSGIW]   [Parse stored GPS information on the Web during idle time]

[GRQI]    [Get request from the Internet PC about parsing of stored GPS information received from the GPRS/3G module during idle time]

[SDI]     [Store parsing data of stored GPS information during idle time]

[SSGIS]   [Send stored GPS information as a text SMS message to the controller user during idle time]

*5.3.2 Interactions Between Different Actors of the Model*
     *IEP-GPS without a Microcontroller*

The interactions between the different actors of the conceived model IEP-GPS without a microcontroller are described in the sequence diagram shown in Figure 5.4. The notations used in Figure 5.4 are as follows:

PCL       Computer PC (Lua)

GPSM      GPS module

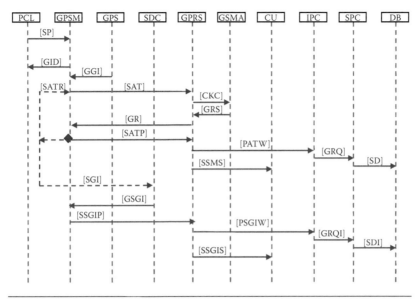

**Figure 5.4**   Interactions between the different actors of the model without a microcontroller.

| | |
|---|---|
| GPS | GPS |
| SDC | SD card |
| GPRS | GPRS/3G module |
| GSMA | GSM antenna |
| CU | Controller user |
| IPC | Internet PC to access the Webpage |
| SPC | Server (PC) |
| DB | Database client/server |
| [SP] | [Send program to configure the microcontroller or for initiation] |
| [GID] | [Get information about program's debugging for display] |
| [GGI] | [Get GPS information] |
| [SAT] | [Send AT command to check the network connectivity] |
| [CKC] | [Check the network connectivity] |
| [GRS] | [Get GPS response signal from the GSM antenna about network connectivity] |
| [GR] | [Get response about network connectivity] |
| [SATP] | [If response is positive, send AT command to be parsed on the Web or via SMS] |
| [PATW] | [Parse AT command over the Web] |

[GRQ]     [Get request from the Internet PC about parsing of information received from the GPRS/3G module]

[SD]     [Store parsing data]

[SSMS]     [Send text SMS message to the controller user]

[SATR]     [Repeat sending AT command to check the network connectivity]

[SGI]     [If no network connection, send GPS information for storage in the SD card]

[GSGI]     [If network connection is back, get stored GPS information back from the SD card]

[SSGIP]     [Send stored GPS information to be parsed on the Web or via SMS]

[PSGIW]     [Parse stored GPS information on the Web during idle time]

[GRQI]     [Get request from the Internet PC about parsing of stored GPS information received from the GPRS/3G module during idle time]

[SDI]     [Store parsing data of stored GPS information during idle time]

[SSGIS]     [Send stored GPS information as text SMS message to the controller user during idle time]

### 5.4 Composition of the Model IEP-GPS

Our proposed model IEP-GPS is composed of different modules. Among these modules, we developed three of them:

- Module 1—Obtaining a location and posting to the Web
- Module 2—Sending SMS as an alert message
- Module 3—Sending a photo by e-mail using the SMTP protocol service

These three modules are each represented by a graph composed of a certain number of phases, as well as the results issued from each phase. These modules are developed in the next chapter as simulations in different programming languages to validate the performance of our proposed model.

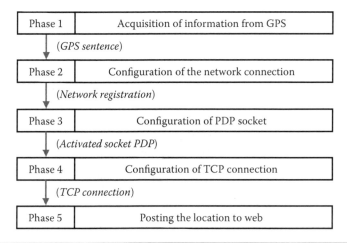

**Figure 5.5**   Module for obtaining a location and parsing to the Web.

### 5.4.1 Module 1—Obtaining a Location and Posting to the Web

Module 1 is composed of five phases, as shown in Figure 5.5, representing the main processes and the flow of information between these processes to obtain a location and posting it to the Web.

1. Phase 1—Acquisition of information from the GPS: This is represented by four steps to get the information from the GPS in order to obtain the GPS sentence in NMEA sentence format (cf. Chapter 4, Section 4.6):
   a. Start GPS session.
   b. Configure port for NMEA sentence.
   c. Get the GPS sentence.
   d. Stop GPS session.
2. Phase 2—Configuration of the network connection: This is represented by a step to do the network registration:
   a. Set the network registration.
3. Phase 3—Configuration of PDP socket: This is represented by three steps to activate the PDP socket:
   a. Set the socket PDP.
   b. Set the authentication (i.e., username and password) for the PDP-IP connections of the socket.
   c. Activate the socket PDP.

4. Phase 4—Configuration of TCP connection: This is represented by a step to establish the TCP connection:
   a. Establish the TCP connection.
5. Phase 5—Posting the location to the Web: This is represented by three steps to send the location represented by the GPS sentence to the Web using TCP service in order to be written in the database implemented on the Server (PC):
   a. Send the TCP message containing the GPS sentence to the TCP server.
   b. Write the TCP message in the database existing on the TCP server "Server (PC)."
   c. Close the socket PDP in order to close all client connections connected with the TCP server.

### 5.4.2 Module 2—Sending SMS as an Alert Message

Module 2 is composed of six phases, as shown in Figure 5.6, representing the main processes and the flow of information between

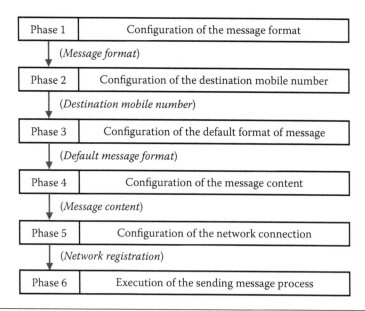

**Figure 5.6** Module for sending SMS as an alert message.

these processes to send a text SMS message to a predefined destination mobile number.

1. Phase 1—Configuration of the message format: This is represented by a step to define the message format:
   a. Set the text SMS message format to text mode.
2. Phase 2—Configuration of the destination mobile number: This is represented by a step to define the destination mobile number receiving the text SMS message sent from a given tracking user:
   a. Set the mobile number of the destination receiving the text SMS message.
3. Phase 3—Configuration of the default format of a message: This is represented by a step to define the default format of the text SMS message:
   a. Set the default format of the text SMS message as an alphabet.
4. Phase 4—Configuration of the message content: This is represented by a step to describe the content of the text SMS message:
   a. Set the content of the text SMS message, where the content is defined according to a coding list defining the type of alert message. For example, the alert message when user1 is moving out of the geofence has two parameters: the username of the tracking unit "user1" and the status "moving out of geofence" depending on predefined coordinates related to the boundaries of the secured areas.
5. Phase 5—Configuration of the network connection: This is represented by two steps to do the network registration:
   a. Set the network registration.
   b. Activate the module GPRS/3G module.
6. Phase 6—Execution of the sending message process: This is represented by a step to send the text SMS message to the destination mobile number:
   a. Send the text SMS message to the predefined destination mobile number.

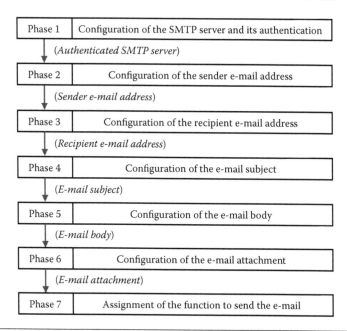

| Phase 1 | Configuration of the SMTP server and its authentication |
| --- | --- |

*(Authenticated SMTP server)*

| Phase 2 | Configuration of the sender e-mail address |
| --- | --- |

*(Sender e-mail address)*

| Phase 3 | Configuration of the recipient e-mail address |
| --- | --- |

*(Recipient e-mail address)*

| Phase 4 | Configuration of the e-mail subject |
| --- | --- |

*(E-mail subject)*

| Phase 5 | Configuration of the e-mail body |
| --- | --- |

*(E-mail body)*

| Phase 6 | Configuration of the e-mail attachment |
| --- | --- |

*(E-mail attachment)*

| Phase 7 | Assignment of the function to send the e-mail |
| --- | --- |

**Figure 5.7**    Module for sending photos by e-mail using a simple SMTP protocol.

### 5.4.3 Module 3—Sending a Photo by E-mail Using the SMTP Protocol Service

Module 3 is composed of seven phases, as shown in Figure 5.7, representing the main processes and the flow of information between these processes to send a snapshot photo by e-mail using the Simple Mail Transfer Protocol (SMTP) service. This e-mail is sent as an alert from a given tracking user's e-mail address to a predefined recipient's e-mail address when this tracking user is out of the geofence (i.e., secured area).

1. Phase 1—Configuration of the SMTP server and its authentication: This is represented by four steps to obtain an authenticated SMTP server:
   a. Set the name of the SMTP server.
   b. Set the value of the port.
   c. Set the username value.
   d. Set the password of the username.
2. Phase 2—Configuration of the sender's e-mail address: This is represented by a step to define the sender's e-mail address of a given tracking user:
   a. Set the e-mail address information of the sender.

3. Phase 3—Configuration of the recipient's e-mail address: This is represented by three steps to define the e-mail address information of the normal recipient as mandatory, and the ones for the Carbon Copy recipient and the Blind Carbon Copy recipient as optional:
   a. Set the e-mail address information of the normal recipient.
   b. Set the e-mail address information of the Carbon Copy recipient.
   c. Set the e-mail address information of the Blind Carbon Copy recipient.
4. Phase 4—Configuration of the e-mail subject: This is represented by a step to define the subject of the e-mail:
   a. Set the subject information of the e-mail, where the content is defined according to a given tracking user when he is moving out of the geofence.
5. Phase 5—Configuration of the e-mail body: This is represented by a step to describe the content of the body of the e-mail:
   a. Set the body information of the e-mail, where the content is defined according to the tracking user when he is moving out of the geofence.
6. Phase 6—Configuration of the e-mail attachment: This is represented by a step to define the attachment of the e-mail that contains the snapshot photo taken when the given tracking user is moving out of the geofence:
   a. Set the attachment file name information.
7. Phase 7—Assignment of the function to send the e-mail: This is represented by a step to send the e-mail containing the snapshot photo to the predefined recipient:
   a. Set the function to send the e-mail.

## 5.5 Conclusion

In this chapter we developed our model InfElecPhy GPS Unit (IEP-GPS) tracking system based on the dedicated 3G/GPRS module for fleet management. This model ensures accurate tracking information. We have introduced the usage of a SD card to store the GPS information in case of disconnection from the network to avoid any loss of data. By introducing the idea of installing the application on the local server of the company, our proposed model allows speeding up the access to collect the required data and solve the limitations of Internet usage. The adaptability of our model with the open source applications, such as OpenGTS, shows the flexibility of our model and its performance (i.e., quality and cost).

In the next chapter, we will present the implementation of the proposed modules of our model IEP-GPS and the performed simulations.

# 6

# APPLICATION

## 6.1 Introduction

In this chapter, we present three samples of application code in order to use a location and parse it to the Web using the TCP service, sending text SMS messages, and sending photos by e-mail using the SMTP service. These three application codes are each written in C programming language, Lua scripting programming language, and the AT command set. The interest in these application codes is to prove the performance of our proposed model (cf. Chapter 5, Section 5.2). We based these application codes on the references offered by Libelium (Libelium 2017) and other technical reports (Boyce 2002; Cormack 2013; Dalakov 1999; Fresh2Refresh 2017; Gao 2013; Nehab 2007; Lua 2011; Tutorialpoints 2017). A simulation of the tracking of a given user is established by using the open source application, OpenGTS (Geotelematic 2015), to display the travel path of this user on Google Maps.

## 6.2 Obtaining a Location and Parsing to the Web

In this section, we present the application code in C, Lua, and the AT command set to obtain a location from GPS in GPRMC and NMEA sentence formats (cf. Chapter 4, Section 4.6), and post to the Web using the TCP service.

### 6.2.1 Code in C

The GPS code and the TCP code in C to send the location in GPRMC and NMEA sentence formats are represented by the code presented in Figures 6.1 through 6.7. This code consists of starting the GPS,

```
const char pin[ ] = "2";
const char apn[ ] = "internet.isp.xx";
const char username[ ] = "internetuser";
const char password[ ] = "password";
const char server[ ] = "hostserver.com";
const char port[ ] = "8080";
const char IP[ ] = "192.168.100.1" ;
int8_t answer;
int onModulePin= 2,aux;
int counter;
char gps_data[100];
char latitude[12],longitude[13];
char date[7],UTC_time[9];
char speed_OG[7],altitude[7];
int content_file=0;
int x=0,y=0;
String N,W,S;
String result;
int data_size = 0;
int end_file = 0;
char aux_str[100];
char data[250];
long previous;
```

**Figure 6.1**   Code in C to do the initialization of the parameters.

```
void setup(){
    pinMode(onModulePin, OUTPUT);
    Serial.begin(115200);
    Serial.println("Starting...");
    power_on();
    delay(5000);
    while( (sendATcommand("AT+CREG?", "+CREG: 0,1", 500) ||
    sendATcommand("AT+CREG?", "+CREG: 0,5", 500)) == 0 );

    // Set APN, username and password
    snprintf(aux_str, sizeof(aux_str), "AT+CGSOCKCONT=1,\"IP\",\"%s\"", apn);
    sendATcommand(aux_str, "OK", 2000);

    snprintf(aux_str, sizeof(aux_str), "AT+CSOCKAUTH=1,1,\"%s\",\"%s\"", username,
    password);
    sendATcommand(aux_str, "OK", 2000);
}
```

**Figure 6.2**   Code in C to set the configuration of the pin on the microcontroller, the baud rate, the server configuration, and its authentication.

getting the GPS sentence, parsing this GPS sentence to GPRMC sentence format or to NMEA sentence format, configure the TCP network, then sending the GPS sentence to the TCP server to be written in the appropriate database using the TCP service.

```
answer = sendATcommand("AT+CGPS=1,1","OK",1000);
  if (answer == 0)
  {
     Serial.println("Error starting the GPS");
     Serial.println("The code stucks here!!");
     while(1);
  }
}
void loop(){
  // Request info from GPS
  answer = sendATcommand("AT+CGPSINFO","+CGPSINFO:",1000);
  if (answer == 1)
  {
     counter = 0;
     do{
        while(Serial.available() == 0);
        gps_data[counter] = Serial.read();
        counter++;
     }
     while(gps_data[counter - 1] != '\r');
     gps_data[counter] = '\0';
     if(gps_data[0] == ',')
     {
        Serial.println("No GPS data available");
     }
     else
     {
```

**Figure 6.3**  Code in C to start a GPS session in standalone mode and get information from the GPS.

### 6.2.2 Code in Lua

The GPS library and the TCP library in Lua to send the location in GPRMC and NMEA sentence formats are represented by the scripting code divided into subscripting codes illustrated in Figures 6.8 through 6.13. These scripting codes consist of starting the GPS, getting the GPS sentence, parsing this GPS sentence to GPRMC sentence format or to NMEA sentence format, configure the TCP network, then sending the GPS sentence to the TCP server to be written in the appropriate database using the TCP service. At each major step of the code, we print trace information on an external AT interface by using the HyperTerminal software (Boyce 2002).

```
        counter = 0;
         y=0;
        do{
           latitude[y]=gps_data[counter];
           y++;
           counter++;
        }while(gps_data[counter]!=',');
        latitude[y]='\0';
        counter++;
        Serial.print(" ");
        Serial.println(gps_data[counter]);    // North or south
        N=String (gps_data[counter]);
        counter+=2;
        y=0;
        do{
           longitude[y]=gps_data[counter];
           y++;
           counter++;
        }while(gps_data[counter]!=',');
        longitude[y]='\0';
        counter++;
        W=String (gps_data[counter]); // West or east
        counter+=2;
        y=0;
        do{
           date[y]=gps_data[counter]; // Date
           y++;
           counter++;
        }while(gps_data[counter]!=',');
        counter++;
        y=0;
        do{
           UTC_time[y]=gps_data[counter]; // UTC time
           y++;
           counter++;
        }while(gps_data[counter]!=',');
        counter++;
        y=0;
        do{
           altitude[y]=gps_data[counter]; // Altitude
           y++;
           counter++;
        }while(gps_data[counter]!=',');
        S="";
        counter++;
        y=0;
        do{
           speed_OG[y]=gps_data[counter]; // Show speed over the ground
           if (gps_data[counter]==','){
             S=S + String ("%2C");
           }
           else {
             S=S + String (gps_data[counter]) ;
           }
           y++;
           counter++;
        }while(gps_data[counter]!=0x0D);
    }
    }
    else
    {
       Serial.println("Error");
    }
    delay(1000);
    sendit();
}
```

**Figure 6.4**   Code in C to extract the GPS data.

```
void sendit(){
TCP_message=(String("acct=test01&dev=test01&code=0xF020&gprmc=%24GPRMC%2C")+String(UTC
_time)+String("%2C")+String("A%2C")+String(latitude)+String("%2C")+String(N)+String("%2C")+String(
longitude)+String("%2C")+String(W)+String("%2C0%2C0%2C")+String(date)+String("%2C")+String(altit
ude)+String("%2C")+String(S));
 Serial.println(TCP_message);
 content_file= TCP_message.length();
 sprintf(aux_str, "AT+NETOPEN=\"TCP\",%s", port);
 answer = sendATcommand(aux_str, "Network opened", 20000);
 if (answer == 1)
 {
  Serial.println("Network opened");
  sprintf(aux_str, "AT+TCPCONNECT=\"%s\",%s", server, port);
  answer = sendATcommand(aux_str, "Connect ok", 20000);
  if (answer == 1)
  {
   Serial.println("Socket opened");
   // Send message to the server
   sprintf(aux_str, "AT+TCPWRITE=%d", strlen(TCP_message));
   answer = sendATcommand(aux_str, ">", 20000);
   if (answer == 1)
   {
    sendATcommand(TCP_message, "Send OK", 20000);
   }
   sendATcommand("AT+NETCLOSE", "OK", 20000);
  }
  else
  {
   Serial.println("Error opening the socket");
  }
 }
 else
 {
  Serial.println("Error opening the network");
 }
}
```

**Figure 6.5**  Code in C to prepare the TCP message of the GPS data in the GPRMC sentence format, connect to the TCP server, and send the TCP message.

### 6.2.3 Code in the AT Command

The AT command is set to obtain the location and send it to parse it to the Web using the TCP service as in Figure 6.14. These AT command sets consist of starting the GPS, getting the GPS sentence in NMEA sentence format, which contains the location of the given user. Then, configure the TCP server and send the GPS sentence using the TCP service to be written in the database on the TCP server. At each step of the code, we have a response display on an external AT interface by using the HyperTerminal software (Boyce 2002). In case the response is OK, the given AT command is successful, otherwise we have to check the error message displayed on the external AT interface.

```
void power_on(){
  uint8_t answer=0;
  // Check if the module is started
  answer = sendATcommand("AT", "OK", 2000);
  if (answer == 0)
  {
    // Power on pulse
    digitalWrite(onModulePin,HIGH);
    delay(3000);
    digitalWrite(onModulePin,LOW);
    Serial.println("");
    Serial.println("");
    Serial.write(0x1A);
    Serial.write(0x00);
    // Wait for an answer from the module
    while(answer == 0){
    // Send AT every two seconds and wait for the answer
     answer = sendATcommand("AT", "OK", 2000);
    }
  }
}
int8_t sendATcommand(char* ATcommand, char* expected_answer1, unsigned int timeout)
{
  uint8_t x=0,  answer=0;
  char response[100];
  unsigned long previous;
  memset(response, '\0', 100);   // Initialize the string
  delay(100);
  while( Serial.available() > 0) Serial.read();   // Clean the input buffer
  Serial.println(ATcommand);   // Send the AT command
  x = 0;
  previous = millis();
  // This loop waits for the answer
  do{
    if(Serial.available() != 0){
      response[x] = Serial.read();
      x++;
      // Check if the desired answer is in the response of the module
      if (strstr(response, expected_answer1) != NULL)
      {
        answer = 1;
      }
    }
    // Wait for the answer with time out
  }
  while((answer == 0) && ((millis() - previous) < timeout));
  return answer;
}
```

**Figure 6.6**   Code in C to check the functionality of the GPRS/3G module to be ready to send the AT Command.

```
void loop(){
  // Request info from GPS
  answer = sendATcommand("AT+CGPSINFO","+CGPSINFO:",1000);
  if (answer == 1)
  {
    counter = 0;
    do{
      while(Serial.available() == 0);
      gps_data[counter] = Serial.read();
      if(gps_data[counter] != '\r' && gps_data[counter] != '\0')
        {
        if(gps_data[counter] == ',')
          {
            gps_datas= gps_datas + String("%2C");
          }
          else
          if(gps_data[counter] == '*')
          {
            gps_datas= gps_datas + String("%2A");
          }
          else
          {
            gps_datas= gps_datas + String(gps_data[counter]);
          }
        }
        else
        {
          if(gps_data[counter] != '\0')
          {
            Serial.print("GPS datas:");
            Serial.println(gps_datas);
            Serial.println("");
            gps_datas = "";
          }
        }
        counter++;
    }
    while(gps_data[counter - 1] != '\r');
    gps_data[counter] = '\0';
    if(gps_data[0] == ',')
    {
      Serial.println("No GPS data available");
    }
    else
    {
      Serial.print("GPS data:");
      Serial.println(gps_data);
      Serial.println("");
    }
  }
  else
  {
    Serial.println("Error");
  }
  delay(5000);
}
```

**Figure 6.7**   Code in C to request information from GPS and convert it to a NMEA sentence.

```
Test_count = 60;
printdir(1)
start = gps.start(1) -- Start the GPS
if(start == 1) then
  print("-----GPS start success!-----\r\n")
end;
count = 0;
while (count < test_count) do
  count = count+1;
  print("-----run count=",count,"-----\r\n");
  str = gps.gpsinfo(); --Get GPS sentence
  print(str,"\r\n","\r\n");
  vmsleep(1000);
end;
close = gps.close();
if(close == 1) then
  print("----GPS close success!----\r\n");
end;
```

**Figure 6.8**  Code in Lua to start the GPS and get the GPS data.

### 6.3 Sending SMS

In this section, we present the application code in C, Lua, and the AT command set to send an alert text SMS message from an originating mobile number set on the GPRS/3G module to a predefined destination's mobile number. The objective of this alert text SMS message is to inform the concerned destination user when the originated user is moving out of the geofence for security purposes.

#### 6.3.1 Code in C

The SMS code in C to send a short text SMS message is presented in Figures 6.15 and 6.16. This code consists of defining the type of the message to be sent and the content of the message for the short text SMS message. The text SMS message will be sent from the originating mobile number set on the GPRS/3G module to a predefined destination's mobile number.

#### 6.3.2 Code in Lua

The SMS library in Lua to send a short text SMS message and a long text SMS message divided into two text SMS messages is represented

```
-- GPS Parsing Module - GPRMC NMEA decoding
local bit = require "bit"
local os = require "os"
local package = require "package"
local stdnse = require "stdnse"
local string = require "string"
_ENV = stdnse.module("gps", stdnse.seeall)
NMEA = {
-- Parser for the GPRMC sentence
  GPRMC = {
    parse = function(str) -- str is the GPS sentence
    local time, status, latitude, ns_indicator, longitude, ew_indicator, speed, course, date,
    variation, ew_variation, checksum =
    str:match("^%$GPRMC,([^,]*),([^,]*),([^,]*),([^,]*),([^,]*),([^,]*),([^,]*),([^,]*),([^,]*),
    ([^,]*), ([^%*]*)(.*)$")
    if ( not(latitude) or not(longitude) ) then
      return
    end
    local deg, min = latitude:match("^(..)(.*)$")
    if ( not(deg) or not(min) ) then
      return
    end
    latitude = tonumber(deg) + (tonumber(min)/60)
    deg, min = longitude:match("^(..)(.*)$")
    if ( not(deg) or not(min) ) then
      return
    end
    longitude = tonumber(deg) + (tonumber(min)/60)
    if ( ew_indicator == 'W' ) then
      longitude = -longitude
    end
    if ( ns_indicator == 'S' ) then
      latitude = -latitude
    end
    return { time = time, status = status, latitude = latitude,
    longitude = longitude, speed = speed, course = course,
    date = date, variation = variation, ew_variation = ew_variation }
  end,
  },
```

**Figure 6.9**   Code in Lua to parse the GPRMC sentence.

by the scripting code divided into subscripting codes illustrated in Figure 6.17. This scripting code consists of defining the type of the message to be sent. That is, the content of the message for the short text SMS message; and, for the long text SMS message, the total number of messages for the sent text message (i.e., 2 in our case). And then there is the sequential SMS number (i.e., 1 for the first message and 2 for the second message) added to the type of the message to be sent and the content of the message for the long text SMS message

```lua
-- Calculate and verify the message checksum
checksum = function(str)    -- str represents the GPS sentence
  local val = 0
  for c in str:sub(2,-4):gmatch(".") do
    val = bit.bxor(val, string.byte(c))
  end
  if ( str:sub(-2):upper() ~= stdnse.tohex(string.char(val)):upper() ) then
    return false, ("Failed to verify checksum (got: %s; expected:
    %s)"):format(stdnse.tohex(string.char(val)), str:sub(-2))
    -- Return false if checksum does not match and errorparsing string if status is false
  end
  return true -- Return status true on success
end,
```

**Figure 6.10**   Code in Lua to calculate and verify the message checksum.

```lua
parse = function(str) -- str contains the GPS sentence
-- Return entry table containing the parsed response or errorparsing string if status is false
-- Return status true on success, false on failure
  local status, errorparsing = NMEA.checksum(str)
  if ( not(status) ) then
  return false, errorparsing
  end
  local prefix = str:match("^%$GP([^,]*)")
  if ( not(prefix) ) then
  return false, "Not a NMEA sentence"
  end
  if ( NMEA[prefix] and NMEA[prefix].parse ) then
    local e = NMEA[prefix].parse(str)
    if (not(e)) then
      return false, ("Failed to parse entry: %s"):format(str)
    end
    return true, e
  else
    local errorparsing = ("No parser for prefix: %s"):format(prefix)
    stdnse.print_debug(2, errorparsing)
    return false, errorparsing
  end
end
}
Util = {
  convertTime = function(date, time)
  local d = {}
  d.hour, d.min, d.sec = time:match("(..)(..)(..)")
  d.day, d.month, d.year = date:match("(..)(..)(..)")
  d.year = d.year + 2000
  return os.time(d)
  end
}
return _ENV;
```

**Figure 6.11**   Code in Lua to parse a GPS sentence using the appropriate parser.

```
TCP_message = "$GPRMC,time,status,latitude,N,longitude,W,speed,course,date,,variation,
ew_variation,checksum";

server="hostserver.com";
port="8080";
IP="192.168.100.1";
apn = "internet.isp.xx";
username= "internetuser";
password = "password";
aux_str="";

-- Network registration
sio.send("ATEO\r\n");
rsp=sio.recv(5000);
cmd1 = string.format("AT+CREG?", "+CREG: 0,1", 500);
cmd2 = string.format("AT+CREG?", "+CREG: 0,5", 500);
sio.send(cmd1);
cm1=sio.recv(5000);
sio.send(cmd2);
cm2=sio.recv(5000);
while ((cm1 or cm2) == 0 ) do
cmd1 = string.format("AT+CREG?", "+CREG: 0,1", 500);
cmd2 = string.format("AT+CREG?", "+CREG: 0,5", 500);
sio.send(cmd1);
cm1=sio.recv(5000);
sio.send(cmd2);
cm2=sio.recv(5000);
end;
```

**Figure 6.12** Code in Lua to set the TCP message in GPRMC format and network registration.

divided into two text SMS messages. The text SMS message (i.e., short or long) will be sent from the originating mobile number set on the GPRS/3G module to a predefined destination's mobile number. At each major step of the code, we print trace information on an external AT interface by using the HyperTerminal software (Boyce 2002).

### 6.3.3 Code in the AT Command Set

The AT command is set to send a text SMS message as is illustrated in Figure 6.18. These AT command sets consist of sending an alert by a text SMS message when user1 is moving out of the geofence using the Internet Service Provider applied to the GPRS/3G Module. This message contains a short text message of the current location of user1 when he is moving out of the geofence. It will be sent from the originating mobile number set on the GPRS/3G module to a predefined destination's mobile number. At each step of the code, we have a response displayed on an external AT interface by using the

```
-- Set the socket PDP
cmd3 = string.format("AT+CGSOCKCONT=1, IP, apn\r\n");
sio.send(cmd3);
rsp=sio.recv(5000);

-- Set the authentication (username and password) for PDP-IP connections of
socket
cmd4 = string.format("AT+CSOCKAUTH=1,1, username, password\r\n");
sio.send(cmd4);
rsp1=sio.recv(20000);

-- Activate the socket PDP
cmd5 = string.format(aux_str,"AT+NETOPEN=IP,\"%s\", port\r\n");
sio.send(cmd5);
answer=sio.recv(20000);
if (answer == 1) then
   print("Network opened");

   -- Establish TCP connection
   cmd7 = string.format(aux_str,"AT+TCPCONNECT=\"%s\",%s", server, port);
   print("cmd7",cmd7);
   cmd8 = string.format(aux_str,"Connect ok", 20000);
   sio.send(cmd8);
   answer=sio.recv(20000);
   if (answer == 1) then
      print("Socket opened");

      -- Send TCP data represented by TCP_message
      cmd9 = string.format(aux_str,"AT+TCPWRITE=\"%d\"", strlen(TCP_message));
      sio.send(cmd9);
      answer=sio.recv(20000);
      cmd10 = string.format(cmd9,">", 20000);
      sio.send(cmd10);
      answer=sio.recv(20000);
      if (answer == 1) then
         cmd11 = string.format(TCP_message, "Send OK", 20000);
         sio.send(cmd11);
      end;

      -- Close socket to close all client's connections connected with the server
      cmd12 = string.format("AT+NETCLOSE", "OK", 20000);
      sio.send(cmd12);
   else
     print("Error opening the socket");
   end;
else
  print("Error opening the network");
end;
```

**Figure 6.13** Code in Lua to establish the TCP connection and send the TCP message to the server.

```
AT+GPS = 1 -- Start GPS session
AT+GPSSWITCH = 1 – Configure port for NMEA sentence
AT+GPSINFO --Get GPS sentence
AT+GPS = 1 -- Stop GPS session

-- Network registration
AT+CREG = 1

-- Set the socket PDP
AT+CGSOCKCONT=1,"IP","internet.isp.xx"

-- Set the authentication (username and password) for PDP-IP connections of socket
AT+CSOCKAUTH=1,1, "internetuser", "password"

-- Activate the socket PDP
AT+NETOPEN="TCP",80

-- Establish TCP connection
AT+TCPCONNECT= "192.168.100.1",80

-- Send TCP to the TCP server
AT+TCPWRITE=55 -- 55 represents the length of GPS sentence

-- Close socket to close all client's connections connected with the server
AT+NETCLOSE
```

**Figure 6.14**   AT Command sets to get the GPS information in NMEA sentence format and send it to the TCP server.

HyperTerminal software (Boyce 2002). In case the response is OK, the given AT command is successful, otherwise we have to check the error message displayed on the external AT interface.

## 6.4  Sending a Photo by E-Mail Using the Simple Mail Transfer Protocol (SMTP) Service

In this section we present the application code in C, Lua, and the AT command set to send a photo taken when the given user is moving out of the geofence. This snapshot photo is sent by e-mail for security purposes using the SMTP service.

### 6.4.1  Code in C

The SMTP code in C to send an e-mail that contains a photo is illustrated in Figures 6.19 through 6.24. This code consists of sending an alert by e-mail when user1 is out of the geofence using the SMTP

```
int8_t answer;
int onModulePin= 2;
char aux_str[30];
//Set the data
const char pin[]="2";
const char phonenumber[]="4080000000";
const char smstext[]="user1 is moving out of geofence";
void setup(){
 pinMode(onModulePin, OUTPUT);
 Serial.begin(115200);
 Serial.println("Starting...");
 power_on();
 delay(3000);
 //Set the PIN Code
 sprintf(aux_str, "AT+CPIN=%s", pin);
 sendATcommand(aux_str, "OK", 2000);
 delay(3000);
 Serial.println("Connecting to the network...");
 while( (sendATcommand("AT+CREG?", "+CREG: 0,1", 500) ||
  sendATcommand("AT+CREG?", "+CREG: 0,5", 500)) == 0 );
 Serial.print("Setting SMS mode...");
 sendATcommand("AT+CMGF=1", "OK", 1000);   // Set the SMS mode to text
 Serial.println("Sending SMS");
 sprintf(aux_str,"AT+CMGS=\"%s\"", phonenumber);
 answer = sendATcommand(aux_str, ">", 2000);   // Send the SMS number
 if (answer == 1)
 {
  Serial.println(smstext);
  Serial.write(0x1A);
  answer = sendATcommand("", "OK", 20000);
  if (answer == 1)
  {
   Serial.print("Sent ");
  }
  else
  {
   Serial.print("error ");
  }
 }
 else
 {
  Serial.print("error ");
  Serial.println(answer, DEC);
 }
}
```

**Figure 6.15**  Code in C to connect to the GSM network and to send a text message.

```
void loop(){
}

void power_on(){
  uint8_t answer=0;
  // Check if the module is started
  answer = sendATcommand("AT", "OK", 2000);
  if (answer == 0)
  {
    // Power on pulse
    digitalWrite(onModulePin,HIGH);
    delay(3000);
    digitalWrite(onModulePin,LOW);
    // Wait for an answer from the module
    while(answer == 0){   // Send AT every two seconds and wait for the answer
      answer = sendATcommand("AT", "OK", 2000);
    }
  }
}
int8_t sendATcommand(char* ATcommand, char* expected_answer, unsigned int timeout){
  uint8_t x=0, answer=0;
  char response[100];
  unsigned long previous;
  memset(response, '\0', 100);   // Initialize the string
  delay(100);
  while( Serial.available() > 0) Serial.read();   // Clean the input buffer
  Serial.println(ATcommand);   // Send the AT command
  x = 0;
  previous = millis();
  // This loop waits for the answer
  do{
  // If there are data in the UART input buffer, read it and check for the answer
    if(Serial.available() != 0){
      response[x] = Serial.read();
      x++;
      // Check if the desired answer is in the response of the module
      if (strstr(response, expected_answer) != NULL)
      {
        answer = 1;
      }
    }
  // Wait for the answer with timeout
  }
  while((answer == 0) && ((millis() - previous) < timeout));
  return answer;
}
```

**Figure 6.16** Code in C to check the functionality of the GPRS/3G module to be ready to send the AT Command.

service. This e-mail contains a snapshot photo of the current location of user1 taken when he is moving out of the geofence. It will be sent from the e-mail address of user1 to the recipient user2, and as optional in Carbon Copy to user2 and in Blind Carbon Copy to user3.

```
function send_sms()
 local sms_content, rst, suc, sms_error_reason;
 local sms_ref, total_sms, seq_num;
 -- Send short text SMS message
 print("send single IRA sms\r\n");
 local phonenumber = "4080000000";
 os.set_cscs(CSCS_IRA); -- Type of message is text sms
 sms.set_cmgf(1); -- Send text mode sms
 sms.set_csmp(17, 14, 0, 0);
 sms_content = "user1 is moving out of geofence"
 suc, sms_error_reason = sms.send(phonenumber, sms_content);--send single sms, default is
"UNSENT"
 print("sms.send=", suc, ",", sms_error_reason, "\r\n");
 -- Send long text sms message
 print("send long IRA sms\r\n");
 os.set_cscs(CSCS_IRA); -- Type of message is text sms
 sms_ref = sms.get_next_sms_ref();
 total_sms = 2; -- Total number of messages for the sent text message = 2
 seq_num = 1; -- Sequential sms number 1
 sms_content = "user1 is moving(1/2)";
 suc, sms_error_reason = sms.send(phonenumber, sms_content, sms_ref, seq_num, total_sms);--
send long sms
 print("sms.send=", suc, ",", sms_error_reason, "\r\n");
 total_sms = 2; -- Total number of messages for the sent text message = 2
 seq_num = 2; -- Sequential sms number 2
 sms_content = "out of geofence(2/2)";
 -- Send long sms
 suc, sms_error_reason = sms.send(phonenumber, sms_content, sms_ref, seq_num, total_sms);
 print("sms.send=", suc, ",", sms_error_reason, "\r\n");

 printdir(1);
 collectgarbage();
 CSCS_IRA=0
 local result;
 result = sms.ready();
 print("sms.ready() = ", result, "\r\n");
 if (not result) then
  print("SMS not ready now\r\n");
  return;
 end;
 send_sms();
```

**Figure 6.17**   Code in Lua to send short and long text SMS messages.

```
AT+CMGF=1 -- Set the SMS message format to text mode
AT+CSCA="+80400000000" -- Set the address through which
 -- mobile originated SMS are transmitted

AT+CSCS=0 -- Format of SMS as default alphabet
AT+CMGS="user1 is moving out of geofence" – Send message
```

**Figure 6.18**   AT Command sets to send text SMS message.

```
int led = 13;          // Set led to digital pin 13
int onModulePin = 2;        // The pin 2 to switch on the module (without press on button)
int x;                 // Define the type of x as integer

//Set server information and its authentication (username and password)
char server[ ]="smtp.smtpservername.com";
char port[ ]="25";
char username[ ]="user1";
char password[ ]="password";

//Set the information about sender, directions and names
char sender_add[ ]="user1@smtpservername.com";
char sender_name[ ]="user1";

//Set the information about normal recipient
char to_add[ ]="user2@gmail.com";
char to_name[ ]="user2";

//Set the information about Carbon Copy recipient
char cc_add[ ]="user3@gmail.com";
char cc_name[ ]="user3";

//Set the information about Blind Carbon Copy recipient
char bcc_add[ ]="user4@gmail.com";
char bcc_name[ ]="user4";

//Set the information about the subject of the e-mail
char subject[ ]="user1 is out of geofence";

//Set the information about the body of the e-mail
char body[ ]="Kindly find attached the snapshot image of the current location
            of user1 when out of geofence";

//Set the information about the attached file to the e-mail
char attachedfile[ ]="C:\\snapshotimage1.jpg";

char response[128];
```

**Figure 6.19** Code in C to set the SMTP server information and the e-mail's required information.

*6.4.2 Code in Lua*

The SMTP library in Lua to send e-mail that contains a photo is represented by the scripting code illustrated in Figure 6.25. This scripting code consists of sending an alert by e-mail when user1 is out of the geofence using SMTP service. This e-mail contains a snapshot photo of the current location of user1 taken when he is moving out of the geofence. It will be sent from the e-mail address of user1 to the recipient user2, and as optional in Carbon Copy to user2 and in Blind Carbon Copy to user3. At each step of the code, we print trace information on an external AT interface by using the HyperTerminal software (Boyce 2002).

```
//Function to switch on the digital pin 2 then switch it off after 2 seconds
void switchModule(){
  digitalWrite(onModulePin,HIGH);   // sets the LED on -- digitalWrite:
  //Write a HIGH value to a digital pin
  delay(2000);   // waits for 2 seconds
  digitalWrite(onModulePin,LOW);   // sets the LED off -- digitalWrite:
  //Write a LOW value to a digital pin
}

//Assign initialization values to parameters
void setup(){
  Serial.begin(115200);   // UART baud rate -- Serial.begin:
  //Set the data rate in bits per second (baud) for serial data transmission
  delay(2000);
  pinMode(led, OUTPUT);            // Configure the digital pin 13 as an OUTPUT
  pinMode(onModulePin, OUTPUT);      // Configure the digital pin 2 as an OUTPUT
  switchModule();              // Switch the module ON
  for (int i=0;i< 5;i++){
    delay(5000);
  }
  delay(5000);
  Serial.println("Connecting to the network..."); // Serial.println:
  // Verification of Network & network registration
  while( (Serial.println("AT+CREG?", "+CREG: 0,1", 500) || Serial.println("AT+CREG?",
  "+CREG: 0,5", 500)) == 0 );

  // Configure the APN of the 3G Sim Card
  sendATcommand("AT+CGSOCKCONT=1,\"IP\",\"internet.isp.xx\"","","OK",1000);

  // Configure the authentication (username & password)
  sendATcommand("AT+CSOCKAUTH=1,1,,\"internetuser\",\"password\"","","OK",1000);

  // Configure the SSL
  Serial.println("AT+CGPSSSL=0","OK",1000); //"AT+CGPSSSL=0" ==> don't use
                                  // certificate; "AT+CGPSSSL=1" ==> use certificate

  answer = Serial.println("AT+CGPS=1,2","OK",1000); // Start GPS session in MS-based mode
  if (answer == 0)
  {
    Serial.println("Error starting the GPS");
    while(1);
  }
}
```

**Figure 6.20**   Code in C to switch on the digital pin and to configure the Internet network.

### 6.4.3 Code in the AT Command Set

The AT command is set to send e-mail that contains a photo as illustrated in Figure 6.26. These AT command sets consist of sending an alert by e-mail when user1 is out of the geofence using the SMTP service. This e-mail contains a snapshot photo of the current location of user1 taken when he is moving out of the geofence. It will be sent

```
void loop(){
digitalWrite(led, HIGH);  // Turn the LED on (HIGH is the voltage level)
 delay(1000);             // Wait for a second
 digitalWrite(led, LOW);   // Turn the LED off by making the voltage LOW
 delay(1000);             // Wait for a second
  // Define the server address and the serial port
  Serial.print("AT+SMTPSRV=\"smtp.smtpservername.com\"",25);

  // Configure the server name and port
  Serial.print(server);
  Serial.print("\",");
  Serial.println(port);
  Serial.flush();
  x=0;
  do{
    while(Serial.available()==0);
    response[x]=Serial.read();
    x++;
  }
  while(!(response[x-1]=='K'&&response[x-2]=='O'));

  // Configure the username and password
  Serial.print("AT+SMTPAUTH=1,\"");
  Serial.print(username);
  Serial.print("\",\"");
  Serial.print(password);
  Serial.println("\"");
  Serial.flush();
  x=0;
  do{
    while(Serial.available()==0);
    response[x]=Serial.read();
    x++;
  }
  while(!(response[x-1]=='K'&&response[x-2]=='O'));
```

**Figure 6.21**  Code in C to configure the SMTP server information.

from the e-mail address of user1 to the recipient user2, and as optional in Carbon Copy to user2 and in Blind Carbon Copy to user3. At each step of the code, we have a response displayed on an external AT interface by using the HyperTerminal software (Boyce 2002). In case the response is OK, the given AT command is successful, otherwise we have to check the error message displayed on the external AT interface.

```
// Configure the sender address and name
Serial.print("AT+SMTPFROM=\"");
Serial.print(sender_add);
Serial.print("\",\"");
Serial.print(sender_name);
Serial.println("\"");
Serial.flush();
x=0;
do{
   while(Serial.available()==0);
   response[x]=Serial.read();
   x++;
}
while(!(response[x-1]=='K'&&response[x-2]=='O'));

// Configure the normal recipient address and name
Serial.print("AT+SMTPRCPT=0,0,\"");
Serial.print(to_add);
Serial.print("\",\"");
Serial.print(to_name);
Serial.println("\"");
Serial.flush();
x=0;
do{
   while(Serial.available()==0);
   response[x]=Serial.read();
   x++;
}
while(!(response[x-1]=='K'&&response[x-2]=='O'));

// Configure the carbon copy recipient address and name
Serial.print("AT+SMTPRCPT=1,0,\"");
Serial.print(cc_add);
Serial.print("\",\"");
Serial.print(cc_name);
Serial.println("\"");
Serial.flush();
x=0;
do{
   while(Serial.available()==0);
   response[x]=Serial.read();
   x++;
}
while(!(response[x-1]=='K'&&response[x-2]=='O'));

// Configure the Blind Carbon Copy recipient address and name
Serial.print("AT+SMTPRCPT=2,0,\"");
Serial.print(bcc_add);
Serial.print("\",\"");
Serial.print(bcc_name);
Serial.println("\"");
Serial.flush();
x=0;
do{
   while(Serial.available()==0);
   response[x]=Serial.read();
   x++;
}
while(!(response[x-1]=='K'&&response[x-2]=='O'));
```

**Figure 6.22**    Code in C to configure the e-mail's required information.

```
// Configure the subject of the e-mail
Serial.print("AT+SMTPSUB=\"");
Serial.print(subject);
Serial.println("\"");
Serial.flush();
x=0;
do{
    while(Serial.available()==0);
    response[x]=Serial.read();
    x++;
}
while(!(response[x-1]=='K'&&response[x-2]=='O'));

// Configure the body of the e-mail
Serial.print("AT+SMTPBODY=\"");
Serial.print(body);
Serial.println("\"");
Serial.flush();
x=0;
do{
    while(Serial.available()==0);
    response[x]=Serial.read();
    x++;
}
while(!(response[x-1]=='K'&&response[x-2]=='O'));

// Configure the attached file of the e-mail
Serial.print("AT+SMTPFILE=1,\"");
Serial.print(attachedfile);
Serial.println("\"");
Serial.flush();
x=0;
do{
    while(Serial.available()==0);
    response[x]=Serial.read();
    x++;
}
while(!(response[x-1]=='K'&&response[x-2]=='O'));
```

**Figure 6.23**  Code in C to configure the e-mail's required information (continued).

## 6.5  Tracking of a Given User on Google Maps Using the Application OpenGTS

This section aims to present the tracking history of the user1 travelling from the New Museum to Washington Square Park, as mentioned in Figure 6.27, by using the application OpenGTS (Geotelematic 2015). The color of the markers indicates the speed, where the green pushpin is for speed less than 15 mph and the red pushpin marks the stationary object. These pushpins are related to a posting of GPS location, of the device of user1, and in the database existing on the server for

```
// Verification of the ISP Internet Network
  Serial.println("AT+CGSOCKCONT=1,\"IP\",\"internet.isp.xx\"","OK",1000);
  // Authentication of the Internet Network (username and password)
  Serial.println("AT+CSOCKAUTH=1,1,,\"internetuser\",\"password\"","OK",1000);

  // without certificate
  Serial.println("AT+CGPSSSL=0","OK",1000);

  // Open network
  Serial.println("AT+NETOPEN=,,1");
  Serial.flush();
  x=0;
  do{
     while(Serial.available()==0);
     response[x]=Serial.read();
     x++;
  }
  while(!(response[x-1]=='K'&&response[x-2]=='O'));

  Serial.println("AT+SMTPSEND");   // Send the email
  Serial.flush();
  x=0;
  do{
     while(Serial.available()==0);
     response[x]=Serial.read();
     x++;
  }
  while(!(response[x-1]=='K'&&response[x-2]=='O'));
  Serial.print("Sending email...");
  x=0;
  do{
     do{
        digitalWrite(led,HIGH);
     }
     while(Serial.available()==0);
     digitalWrite(led,LOW);
     response[x]=Serial.read();
     x++;
  }
  while(!(response[x-1]=='S'&&response[x-2]=='S'));
  printf("%s",response);
  while(1);
}
```

**Figure 6.24** Code in C to configure the Internet network information and to send the e-mail.

each 5 minutes. This time is predefined in the application OpenGTS. In Figure 6.27, we can see the geofence representing the predefined secured area in the application OpenGTS, as well as that user1 is moving out of this geofence by stopping at Washington Square Park.

```
Local smtp = require("socket.smtp") -- Load the SMTP module and all its requirements
printdir(1) -- Print to external AT interface
collectgarbage(); -- Keeping track of memory usage

--Set the configuration of SMTP server and its authentication (i.e., username and password)
rst = smtp.config("smtp.smtpservername.com",25,"user1","password");
print("rst of smtp.config=", rst, "\r\n"); -- Print trace information to external AT interface

-- Set the configuration of the sender address and name information of the e-mail
rst = smtp.set_from("<user1@smtpservername.com>", "user1");
print("rst of smtp.set_from=", rst, "\r\n");

-- Set the configuration of the recipient address and name information of the e-mail
rst = smtp.set_rcpt(0,0,"<user2@gmail.com>", "user2"); -- To normal recipient
print("rst of smtp.set_rcpt=", rst, "\r\n");
rst = smtp.set_rcpt(0,1,"<user3@gmail.com>", "user3"); -- Carbon Copy recipient
print("rst of smtp.set_rcpt=", rst, "\r\n");
rst = smtp.set_rcpt(0,2,"<user4@gmail.com>", "user4"); -- Blind Carbon Copy recipient
print("rst of smtp.set_rcpt=", rst, "\r\n");

-- Set the configuration of the subject information of the e-mail
rst = smtp.set_subject("user1 is out of geofence");
print("rst of smtp.set_subject=", rst, "\r\n");

-- Set the configuration of the body information of the e-mail
rst = smtp.set_body("Kindly find attached the snapshot image of the current location of user1 when
moving out of geofence.");
print("rst of smtp.set_body=", rst, "\r\n");

-- Set the configuration of the attachment information of the e-mail containing the snapshot
-- image file of the current location of user1 when moving out of geofence
rst = smtp.set_file(1, "c:\\snapshotimage1.jpg");
print("rst of smtp.set_file=", rst, "\r\n");

-- Send the e-mail
print("smtp sending now...\r\n");
rst = smtp.send(); -- function used to send e-mail
print("rst of smtp.send=", rst, "\r\n");
```

**Figure 6.25**    Code in Lua to send an image using the SMTP service.

By clicking with the mouse on this point, we will get the following information:

- Data point: [#12]
- Device name: User1
- Status: stop
- Date: 2017/07/10
- Time: 11:30:20
- GPS Latitude/Longitude: 40.7256335, −73.9929977
- Speed: 0.0 mph
- Address: Washington Square Park, New York, NY, US

```
--Set the configuration of SMTP server address and port number
AT+SMTPSRV = "smtp.smtpservername.com", 25

--Set the configuration of SMTP server authentication (i.e., username and password)
AT+SMTPAUTH = 1, "user1", "password"    -- 1: SMTP server requires authentication
                                        -- 0: SMTP server does not require authentication

-- Set the configuration of the sender address and name information of the e-mail
AT+SMTPFROM = "user1@smtpservername.com", "user1"

-- Set the configuration of the recipient address and name information of the e-mail
AT+SMTPRCPT = 0, 0, "user2@gmail.com", "user2" -- To normal recipient
AT+SMTPRCPT = 1, 0, "user3@gmail.com", "user3" -- Carbon Copy recipient
AT+SMTPRCPT = 2, 0, "user4@gmail.com", "user4" -- Blind Carbon Copy recipient

-- Set the configuration of the subject information of the e-mail
AT+SMTPSUB = "user1 is out of geofence"

-- Set the configuration of the body information of the e-mail
AT+SMTPBODY = "Kindly find attached the snapshot image of the current location of user1 when
moving out of geofence."

-- Set the configuration of the attachment information of the e-mail containing the snapshot
-- image file of the current location of user1 when moving out of geofence
AT+SMTPFILE = 1, "c:\\snapshotimage1.jpg"

-- Send the e-mail
AT+SMTPSEND
```

**Figure 6.26**   AT Command sets to send an image using the SMTP service.

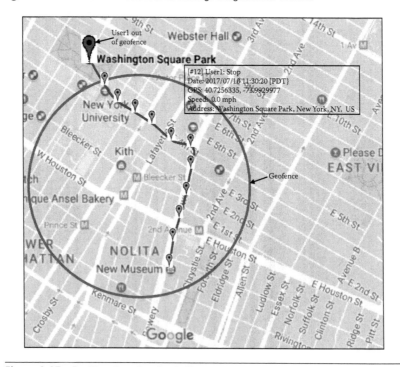

**Figure 6.27**   Tracking of user1 on Google Maps.

## 6.6 Conclusion

In summary, we have presented in this chapter three samples of application code in order to obtain a location and parse it to the Web using the TCP service, sending a text SMS message, and sending a photo by e-mail using the SMTP service. These three application codes are each written in C programming language, Lua scripting programming language, and the AT command set to prove the performance of our proposed model. A simulation of the tracking of a given user is integrated to justify the accuracy of the tracking process presented in our model and its efficiency.

# Conclusion and Perspectives

A study of the extensive bibliography on the field of tracking systems for fleet management to get accurate information shows that our model InfElecPhy GPS Unit (IEP-GPS) is topical and has new features.

This model is based on the dedicated 3G/GPRS module for fleet management and is founded on the use of microcontroller architecture or not using microcontroller architecture to manage the exchange of data with the end users.

This model ensures accurate tracking information for fleet management when disconnection occurs from the network by doing the storage of GPS information on a SD card.

The model IEP-GPS is designed having two architectures (i.e., with microcontroller and without microcontroller) consisting of electronic equipment of different types. These interact in a coherent way along with specific types of communication protocols to manage the operation of the tracking system in a well-organized manner.

The problem of disconnection from the network and the limitation of Internet usage is an important issue in the field of research.

The implementation of our model based on the simulations of three modules has shown us the effectiveness and the performance of our integrated modules.

The usage of the SD card for tracking information storage, when disconnection from a network occurs, has shown in practice its accuracy compared to other systems.

The introduction of the open source application, OpenGTS, to manage the data and edit the required reports is efficient at the managerial level and the costs are less as compared to what has been developed in this field.

Our approach to implementation of a local server in the organization to do the tracking for the fleet management of their users has a proven efficiency at the level of execution time to manage the data and edit the required reports. The implementation also does well regarding the limitations of Internet usage and the time required for execution.

Our method of designing the model with two architectures (i.e., with microcontroller and without microcontroller) has given a flexibility of choice for the clients at the level of cost and the customized size of the device.

Our model should be applied in several fields such as science and fleet management.

Further perspectives consist of:

- Implementing our model in primary schools, homes for the elderly or any other field that requires accurate tracking data.
- Having an extension of our tracking model by the integration of multiple covers and shapes, such as for watches and bracelets. This is in order to have the ability for children or the elderly to wear it only in very restricted conditions, such as when the children are moving out of predefined secured areas.

# References

Advanced Tracking. 2012. Tracking Device GSM or Satellite. http://www.advanced-tracking.com/gsm-or-satellite.htm (accessed April 2, 2017).

Aebi, L. 2007. *Découvrir l'univers du GPS & Exploiter son potentiel*. Libro Veritas, France.

Aerospace. 2010. Military Satellite Communications Fundamentals. In *Crosslink Magazine*, USA. http://www.aerospace.org/crosslinkmag/spring-2010/military-satellite-communications-fundamentals/ (accessed March 30, 2017).

Agarwal, T. 2015. Microcontrollers—Types & Applications. http://www.elprocus.com/microcontrollers-types-and-applications (accessed February 16, 2017).

Agarwal, T. 2016. Difference between AVR, ARM, 8051 and PIC Microcontrollers. https://www.elprocus.com/difference-between-avr-arm-8051-and-pic-microcontroller/ (accessed February 16, 2017).

Agarwal, T. 2017. What is the Simulator and Emulator for 8051 Microcontroller. In *Official Blog*, Edfefx Technologies Pvt Ltd., Kits & Solutions. http://www.edgefxkits.com/blog/8051-microcontroller-simulator-and-emulator/ (accessed April 25, 2017).

Alani, M. M. 2014. *Guide to OSI and TCP/IP Models*. Springer, New York.

AllAboutEE. 2012. How to Program a Microcontroller—What Do I Need? https://www.youtube.com/watch?v=FwBdO-dCd0E (accessed March 12, 2017).

Andred-Sanchez, P., and J. T. Heun 2011. A general guide to global positioning systems (GPS)—Understanding operational factors for agricultural applications in Arizona. In *Furtherance of Cooperative Extension Work*, College of Agriculture and Life Sciences, The University of Arizona, AZ, USA. 1553.

AT&T. 2016. Frequently Asked Questions Regarding 2G Sunset. https://www.business.att.com/content/other/2G-Sunset-FAQ_2016.pdf (accessed March 30, 2017).

Atmel. 2012. AVR1010: Minimizing the Power Consumption of Atmel AVR XMEGA Devices. http://www.atmel.com/images/doc8267.pdf (accessed April 30, 2017).

Atmel. 2016. AVR910: In-system programming. In *Application Note*. http://www.atmel.com/Images/Atmel-0943-In-System-Programming_ApplicationNote_AVR910.pdf (accessed March 3, 2017).

Aughey, R. J. 2011. Applications of GPS technologies to field sports. *International Journal of Sports Physiology and Performance*, 6(3): 295–310.

AuScope. 2014. Application of GPS. AuScope GPS in Schools Project. University of Tasmania in conjunction with Geoscience Australia. http://dpipwe.tas.gov.au/Documents/Worksheet%203%20-%20Applications%20of%20GPS.pdf (accessed January, 12, 2017).

Bilal, M. 2017. Pic Microcontroller Programming in C using Mikroc Pro for PIC. http://microcontrollerslab.com/pic-microcontroller-programming-c/ (accessed April 30, 2017).

Blacksys. 2015. Blacksys CH-100B. Cammsys Blackbox. In *User Manual*. http://www.produktinfo.conrad.com/datenblaetter/1300000-1399999/001359580-an-01-en-BLACKSYS_CH_100B_AUTOKAMERA.pdf (accessed March 30, 217).

Blum, R. 2006. *Professional Assembly Language*. Wiley Publishing, USA.

Boyce, J. 2002. Step-by-step: How to use the HyperTerminal tool to troubleshoot modem problems. In *TechRepublic*, http://www.techrepublic.com/article/step-by-step-how-to-use-the-hyperterminal-tool-to-troubleshoot-modem-problems/ (accessed March 13, 2017).

Brown, T. M. L., S. A. McCabe, and C. Welford. 2007. Global positioning system (GPS) technology for community supervision: Lessons learned. In *Noblis Technical Report*, Noblis, USA.

CameraDecision. 2016. Best Compact Cameras with GPS. http://cameradecision.com/features/Best-Compact-cameras-with-GPS (accessed March 3, 2017).

Cazaubon, C. 1997. *Les microcontrôleurs HC11 et leur programmation*. Masson, Paris.

Chatterjee, A. 2009. Role of GPS in Navigation, Fleet Management and other Location Based Services. https://www.geospatialworld.net/article/role-of-gps-in-navigation-fleet-management-and-other-location-based-services/ (accessed October 20, 2016).

Choudhary, H. 2012. Microcontroller Programmer/Burner. https://www.engineersgarage.com/tutorials/microcontroller-programmer-burner (accessed March 7, 2017).

Christopher, Z. 2014. Some GPS monitoring devices capable of audio recording. In *Prison Legal News*, Electronic Monitoring, Attorney Client, Puerto Rico, p. 52.

Circuit Basics. 2016. Basics of UART Communication. http://www.circuit-basics.com/basics-uart-communication/ (accessed April 3, 2017).

Cockerell, P. J. 1987. *ARM Assembly Language Programming*. MTC, England.

Comer, D. E. 2000. *Internetworking with TCP/IP: Principles, Protocols, and Architectures*. Department of Computer Sciences, Purdue University, West Lafayette, Indiana, USA, Vol I, Fourth Edition. Prentice Hall, New Jersey.

Cormack, J. 2013. Programming a Microcontroller with Lua. http://www.londonlua.org/oshug-presentation/ (accessed May 1, 2017).

Corvallis Microtechnology. 2000. Introduction to the Global Positioning System for GIS and TRAVERSE. http://www.cmtinc.com/gpsbook/ (accessed January 3, 2017).

Dalakov, G. 1999. The Modem of Dennis Hayes and Dale Heatherington. http://history-computer.com/ModernComputer/Basis/modem.html (accessed May 2, 2017).

Das, S. K. 2013. GPS, GIS and Their Uses. General Technical Information. https://shivkumardas.wordpress.com/agri-tech/an-introduction-to-gps-gis-and-its-uses-in-agriculture/ (accessed December 22, 2016).

DePriest, D. 2013. NMEA Data. http://www.gpsinformation.org/dale/nmea.htm (accessed February 20, 2017).

Difference Between. 2017. Difference between GSM and GPS. http://www.differencebetween.info/difference-between-gsm-and-gps (accessed April 2, 2017).

Dunn, M. J. 2013. Global Positioning Systems Directorate. Public release, Systems Engineering & Integration, Interface Specification Is-GPS-200, USA.

Edwards, J., and R. Bramante. 2009. *Networking Self-Teaching Guide: OSI, TCP/IP, LANs, MANs, WANs, Implementation, Management and Maintenance*. Wiley Publishing, USA.

Electrical Engineering. 2012. What is a Bootloader, and How Would I Develop One? http://electronics.stackexchange.com/questions/27486/what-is-a-boot-loader-and-how-would-i-develop-one (accessed January 30, 2017).

Electronics Hub. 2015. Interfacing GPS with 8051 Microcontroller. http://www.electronicshub.org/?s=Interfacing+GPS+with+8051+Microcontroller (accessed March 13, 2017).

Electronics Hub. 2017. Microcontroller types and applications. In *Electronics Tutorial*. http://www.electronicshub.org/microcontrollers (accessed February 15, 2017).

El-Rabbany, A. 2002. *Introduction to GPS: The Global Positioning System*. Artech House, USA.

EngineersGarage. 2012. Bootloader. https://www.engineersgarage.com/tutorials/bootloader-how-to-program-use-bootloader (accessed February 20, 2017).

Esa. 2013. Satellite Frequency Bands. http://www.esa.int/Our_Activities/Telecommunications_Integrated_Applications/Satellite_frequency_bands (accessed April 3, 2017).

Forouzan, B. A. 2000. *TCP/IP Protocol Suite* (First ed.). Tata McGraw-Hill Publishing Company Limited, India.

Fresh2Refresh. 2017. C Programming Tutorial. http://fresh2refresh.com/c-programming/c-language-history/ (accessed April 30, 2017).

FTDI Chip. 2017. FTDI Products. http://www.ftdichip.com/FTProducts.htm (accessed March 7, 2107).

Gao, L. 2013. The.Lua.Tutorial. http://luatut.com/ (accessed February 5, 2017).

Garmin. 2011. Nüvi 30/40/50. In *Owner's Manual*, Taiwan. https://www.gpscity.com/pdfs/manuals/NUVI30-EN-MANUAL.pdf (accessed March 7, 2017).

GCMD. 2008. Definitions of geocentric orbits from the Goddard Space Flight Center. In *Wayback Machine Internet Archive*. https://web.archive.org/web/20100527132541/http://gcmd.nasa.gov/User/suppguide/platforms/orbit.html (accessed March 30, 2017).

Génération Robots. 2017. New 3G + GPS shield for Arduino. In *Technical documentation*. https://www.generationrobots.com/media/3G-GPRS-GPS-Arduino-Shield-With-Audio-Video-Kit.pdf (accessed March 20, 2017).

Geotelematic. 2015. http://www.geotelematic.com/ (accessed January 12, 2016).

Gibilisco, S., and M. Doig. 2017. Machine Code (Machine Language). http://whatis.techtarget.com/definition/machine-code-machine-language (accessed April 30, 2017).

GSMA. 2017. Mobile Technology. http://www.gsma.com/aboutus/gsm-technology (accessed March 3, 2017).

Huurdeman, A. A. 2003. *The World History of Telecommunications*. John Wiley, NJ.

Ierusalimschy, R., L. H. de Figueiredo, and W. C. Filho. 1996. Lua—An extensible extension language. *Software: Practice and Experience (SPE)*, 26 (6): 635–652.

Ierusalimschy, R., L. H. de Figueiredo, and W. C. Filho. 2007. The evolution of Lua. In *The Proceedings of the Third ACM SIGPLAN Conference on History of Programming Languages (HOPL III)*, ACM, USA, 2-1–2-26.

Jan, S. 2010. GPS Segments/Components. xa.yimg.com/kq/groups/21620206/615119600/name/3.+GPS+Segments.ppt (accessed December 19, 2016).

Kamal, R. 2012. *Microcontroller: Architecture, Programming Interfacing and System Design*. Pearson Education, USA.

Khadse, R., N. Gawai, and B. M. Faruk. 2014. Overview and comparative study of different microcontrollers. In *The International Journal for Research in Applied Science & Engineering Technology (IJRASET)*, 311–315, IJRASET 2(XII), USA.

Leadtek. 2012. GPS Protocol Reference Manual. http://www.elgps.com/public_ftp/Documentos/SIRF_Protocol.pdf (accessed March 15, 2017).

Libelium. 2017. www.libelium.com. (Copyright Permission May 22, 2017).

LiveViewGPS. 2007. GPS GSM Tracker. https://www.liveviewgps.com/gps+gsm+tracker+.html (accessed April 2, 2017).

Lua. 2011. About Lua. http://www.lua.org/about.html#why (accessed May 1, 2017).

Mazidi, M. A., R. D. Mckinlay, and D. Causey. 2013. *PIC Microcontroller and Embedded System Using Assembly and C for PIC18*. Pearson Education, USA.

Mazzei, D., G. Montelisciani, and G. Baldi. 2015. Changing the programming paradigm for the embedded in the IoT domain. In *2nd World Forum of the Internet of Things (WF-IoT)*, IEEE, Milan, pp. 239–244.

McNamara, J. 2004. *GPS for Dummies*. Wiley, Indiana.

Motschenbacher, C. D., and J. A. Connelly. 1993. *Low-Noise Electronic System Design*. John Wiley, USA.

Nehab, D. 2007. Network Support for the Lua Language. http://w3.impa.br/~diego/software/luasocket/ (accessed February 5. 2017).

Neilan, R., and J. Kouba, 2000. Introduction to GPS—The IGS—NASA. USA. ftp://igscb.jpl.nasa.gov/pub/resource/tutorial/Intro_GPS.ppt (accessed November 24, 2016).

NewbieHack. 2014. Microcontroller—A Beginners Guide—Introduction to Interrupts—Using the Timer/Counter as an Example. PHD Robotics, LLC, US. https://www.newbiehack.com/IntroductiontoInterrupts.aspx (accessed April 30, 2017).

Novatel. 2010. Applications of High-Precision GNSS. https://www.novatel.com/industries/ (accessed January 13, 2017).

NPTEL. 2009. Module 3 Data Link control. In *Courses of Computer Science and Engineering*, version 2 CSE, Indian Institute of Technology (IIT), Kharagpur, India. http://nptel.ac.in/courses/106105080/pdf/M3L2.pdf (accessed March 3, 2017).

Nursat, S. A. 2010. Introduction to AT Commands and its Uses. https://www.codeproject.com/Articles/85636/Introduction-to-AT-commands-and-its-uses (accessed May 2, 2017).

OpenStreetMap. 2016. Audio Mapping. http://wiki.openstreetmap.org/wiki/Audio_mapping#Possible_problems_with_audio_mapping (accessed March 7, 2017).

Ozden, O. 2013. CISC architecture and RISC architecture. In *Journey Towards Completing a MSc Degree*. http://oozden.wordpress.com/2013/02/08/cisc-architecture-and-risc-architecture (accessed February 16, 2017).

Parai, M. K., B. Das, and G. Das. 2013. An overview of microcontroller unit: From proper selection to specific application. *International Journal of Soft Computing and Engineering*, 2(6): 228–231.

Peatman, J. B. 1988. *Design with Microcontrollers*. McGraw—Hill, USA.

Pololu. 2015. *Pololu Wixel User's Guide*. https://www.pololu.com/docs/0J46 (accessed March 5, 2017).

Raju, P. L. N. 2004. Fundamentals of GPS. In *Proceedings of a Training Workshop on the Satellite Remote Sensing and GIS Applications in Agricultural Meteorology Programme (AgMP)*, Geoinformatics division, Indian Institute of Remote Sensing, ed. Sivakumar, M.V.K., Roy, P.S., Harmsen, K., and Saha, S. K., 1182: 121–150. WMO, India.

RF Wireless World. 2012. GPS vs GPRS. http://www.rfwireless-world.com/Terminology/GPS-vs-GPRS.html (accessed April 2, 2017).

Roberson, G. T. 2005. *GPS Applications in Agriculture. Agricultural Machinery Systems, Biological & Agricultural Engineering*, NC State University, USA. *PrecIsIonAg* 1–33.

Rodgers, S. 2007. *Types of GPS Receivers. Presentation, Engineering Information Services*, UNC Chapel Hill, USA.

Rodriguez, B. 1995. Moving Forth—Part 7: CamelForth for the 8051. *The Computer Journal*, 71.

RoseMary. 2010. 8051 and 8051 Microcontroller. In *Tutorial Atmel 8051 Architecture EngineersGarage*. http://www.engineersgarage.com/8051-microcontroller (accessed February 13, 2017).

Rouse, M. 2008. Satellite. http://searchmobilecomputing.techtarget.com/definition/satellite (accessed March 13, 2017).

Rouse, M. 2016. Global Positioning System (GPS). http://searchmobilecomputing.techtarget.com/definition/Global-Positioning-System (accessed March 13, 2017).

Sauter, M. 2014. *From GSM to LTE-Advanced: An Introduction to Mobile Networks and Mobile Broadband* (Second edn.). John Wiley, UK.

Schurman, K. 2017. The 7 Best GPS Cameras to Buy in 2017. https://www.lifewire.com/best-gps-cameras-493669 (accessed March 7, 2017).

Sharma, S., R. Kumar, and P. Bhadana. 2013. Terrestrial GPS positioning system. *International Journal of Research in Engineering and Technology (IJRET)*, 2(4): 584–589. http://esatjournals.net/ijret/2013v02/i04/IJRET20130204031.pdf (accessed September 16, 2016).

Sheriff, R. E., and Y. Fun Hu. 2001. *Mobile Satellite Communication Networks*. John Wiley, England.

Silverthorne, V. 2016. Integrated Development Environment (IDE). http://searchsoftwarequality.techtarget.com/definition/integrated-development-environment (accessed March 10, 2017).

Singtel. 2015. Singapore to Cease 2G Services from April 2017. http://info.singtel.com/node/14089 (accessed April 1, 2017).

Surratt, D. 2003. Magellan Meridian FAQ. *Updated Info Provided by the Meridian Group Members, version 2.0*. http://gpsinformation.net/MeridianFAQv2_0.pdf (accessed March 7, 2017).

Svendsli, O. J. 2003. Atmel's self-programming flash microcontroller. In *Atmel White Paper*, Atmel Corporation, San Jose, USA, pp. 1–6.

Swarthmore. 2010. Serial communications: RS232, SPR, I2C. In *Lectures E91(10)*, Swarthmore College, USA.

TeleGeography. 2016. Telstra Switches Off Gsm Network. https://www.telegeography.com/products/commsupdate/articles/2016/12/02/telstra-switches-off-gsm-network/ (accessed April 1, 2017).

TTU. 2012. Principles of GPS. In *Lectures Documents*, Geospatial Center, Texas Tech University, Texas, USA. http://www.depts.ttu.edu/geospatial/center/gist4310/documents/lectures/Fall%202012/4310-03%20Principles%20of%20GPS%20I_Trilateration.pdf (accessed September 14, 2016).

Tutorialpoints. 2017. C Language—Overview. https://www.tutorialspoint. com/cprogramming/c_overview.htm (accessed April 30, 2017).

U-blox. 2009. GPS antennas—RF design considerations for u-blox GPS receivers. In *Application Note*. https://www.u-blox.com/sites/default/ files/products/documents/GPS-Antenna_AppNote_%28GPS-X-08014%29. pdf?utm_source=en%2Fimages%2Fdownloads%2FProduct_ Docs%2FGPS_Antennas_ApplicationNote%28GPS-X-08014%29.pdf (accessed February 13, 2017).

Wilson, T. 2016. The Role GPS Devices Play in Vehicle Tracking and Fleet Management. http://trackimocom/gps-devices-in-vehicle-tracking-fleet-management/ (accessed December 15, 2016).

# Index